MATHS'
TRICKS
to Blow Your
MIND

Kyle D Evans is a maths teacher and award-winning maths presenter and entertainer. He regularly gives talks (and sings) about maths at schools, festivals and comedy clubs, and has performed everywhere from Edinburgh Fringe to London's Science Museum.

@kyledevans

/borntosum

www.kyledevans.com

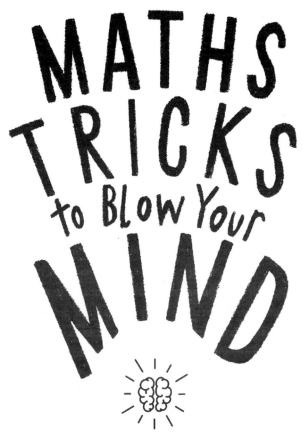

MATHS TRICKS to Blow Your MIND

A Journey Through Viral Maths

KYLE D EVANS

ALLEN&UNWIN

For Edwin and Juno, and all the emergency homeschooling parents.

First published in the United Kingdom in 2021 by Allen & Unwin, an imprint of Atlantic Books Ltd.

This paperback edition first published in 2022

Copyright © Kyle D. Evans, 2021

The moral right of Kyle D. Evans to be identified as the author of this work has been asserted by him in accordance with the Copyright, Designs and Patents Act of 1988.

10 9 8 7 6 5 4 3 2 1

A CIP catalogue record for this book is available from the British Library.

Paperback ISBN: 978 1 83895 876 3
E-book ISBN: 978 1 83895 367 6

Printed in Great Britain by Clays Ltd, Elcograf S.p.A.

Allen & Unwin
An imprint of Atlantic Books Ltd
Ormond House
26–27 Boswell Street
London
WC1N 3JZ

www.allenandunwin.com/uk

CONTENTS

INTRODUCTION

My smartphone lights up with the familiar red blob: I have a Facebook notification. I can see that it's from an old school friend I haven't seen for 20 years; they're probably just inviting all their contacts to 'like' their new business venture. Or, even worse, they could have found an old photo from the class of 2000, when my nickname was 'Spud' because my head looked like a potato. I'm definitely going to ignore it.

But… what's the worst that could happen? Hey, I'm a sucker for that dopamine hit, and, besides, I'm just sitting here doing nothing while my kids watch *Moana* for the third time today. What the hell, I'll have a look:

> Maths trick: A% of B = B% of A.
> So 8% of 25 is equal to 25% of 8.
> 25% of 8 = 2.
>
> Spud! You're into maths right? How did I never know this?!

Yes, I suppose it would be fair to say that I'm 'into maths', spending most of my working life teaching maths in a sixth-form college, then on weekends changing costume in a phone booth to perform live maths shows for children and adults. Yet the number of people I have ever taught or performed to pales in comparison to the number of people who engage with viral social media maths puzzles and 'life hacks' like the above.

It is quite amazing, isn't it? Did you know that 8% of 25 is the same as 25% of 8? I think of this as a wonderful example of the good side of 'viral' maths: something that everyone who's been through school can relate to, and is obvious once you've seen it, but the majority of people seem to go through their entire education without knowing about. What better use of social media? Raising the mathematical literacy of millions of people – who could be cynical about that?*

On the other hand, we have the dark side of viral maths: the provocative, binary choice strand of social media problems that are designed to provoke polarization and division (not that type):

Let's see who's dumb.

$60 + 60 \times 0 + 1 = ?$

1

61

121

This type of problem seems to be the most popular of all viral maths questions, probably because it is intentionally divisive. If we've learnt anything since the invention of the internet, it's that people just can't resist (virtually) screaming at any old stranger on the other side of the world who is naive enough to think that the answer is 1 when it's *obviously* 61!

---

* Inevitably, many people can be cynical about that: more in Chapter 1.

Either way, I'm fascinated by what makes things go viral, and being a jack of all mathematical trades I'm especially interested in why certain maths problems take off while others flounder.

I feel it's important to mention up front that this is a book entirely about 'viral' social media maths, and absolutely in no way about the mathematics of virus transmission. There are many great books about that particular topic and I am barely fit to lace the boots of the good people working in this area. What exactly do we mean when we say something has 'gone viral' though? When did the language of viruses and epidemics start being used to describe popular videos, graphics or even just ideas?

Richard Dawkins is the godfather of the word 'meme', first using it in his 1976 bestseller *The Selfish Gene*, meaning an idea, behaviour or style that spreads from person to person within a culture.* But the online etymology dictionary actually dates the use of 'viral' outside of medicine even earlier, with Jeffrey Rayport of Harvard Business School first using the term 'viral marketing' in 1972. Scratch a little deeper (on Wikipedia) and you'll find the philosopher Marshall McLuhan describing technology as being 'virulent in nature' as far back as 1964. But despite this there can be no doubt that 'viral' has only really taken off as a term separate from actual diseases since the age of social media.

Interestingly, when I asked friends and family across various generations to name what they thought of as the all-time epitome of a viral video, tweet, image or meme, the most popular response to come back was the 'Ellen Oscars selfie'. This was a photograph taken by the television presenter Ellen DeGeneres at the 2014 Oscars, featuring A-listers such as Brad Pitt, Julia Roberts, Meryl Streep, Lupita Nyong'o

---

* It's somewhat unfortunate that Dawkins, having practically created the idea of memes and virality, now writes some of the most tone-deaf, coldest takes in all of social media.

and, fantastically, Lupita's opportunistic non-acting plus-one, her brother Peter. This image was for some time the most retweeted post in Twitter history, though it has since been surpassed by four tweets: two by Japanese billionaire Yusaku Maezawa in which he promised to give away vast sums of money to retweeters, one commemorating the passing of the actor Chadwick Boseman, and one by an American teenager begging the fast-food outlet Wendy's for free chicken nuggets. That was not a paragraph I ever expected I'd type.

Crucially though, I would *not* classify any of the above as truly viral in nature, other than perhaps Carter Wilkerson, the nuggets guy. All the others were broadcast by accounts with an extremely high following, from which almost all shares directly followed. Wilkerson's plea for free 'nuggs', on the other hand, picked up gradual momentum over the course of a month as more and more people saw the humorous story – the David vs Goliath angle of a man attempting to achieve a truly impossible goal* – and began to talk about the story and share it. That's what we all mean by a virus, isn't it? Something that starts slowly and builds remarkably quickly due to fertile conditions for spreading. I'm no epidemiologist, but I don't think there has ever been a biological virus that started by being transmitted simultaneously to millions of people from the same one original source. Admittedly most of my knowledge in this area comes from watching the 2011 movie *Contagion* repeatedly, but I still think I'm correct on that one.

We could consider everything that goes viral on the internet to be on a scale of *structural virality*,[†] where at one end you have 'broadcast'

---

* Wilkerson asked Wendy's how many retweets he needed to achieve for a year of free nuggets, and the reply from Wendy's was '18 million'. For context, the Ellen selfie is currently on around 3 million retweets; nothing has ever surpassed 5 million. Wendy's caved in and gave him the year's supply when he got past 3 million though.

† Goel et al., 'The structural virality of online diffusion', *Management Science*, Vol. 62, 2016.

posts such as the Ellen selfie, where the original poster is linked to people who have seen it largely by one direct link (as in the image on the left). At the other end of the scale (see the image on the right) you have the kind of post that I'll be dealing with in this book: posts that start from a modest source but spread rapidly from person to person, via shares, comments or even word of mouth. Think of recent viral sensations such as the dress that was either white and gold or blue and black, or the sound clip that is either 'Brainstorm' or 'Green Needle'.* Their contentious nature made people want to share them and find out what other people thought, and this naturally drives virality.

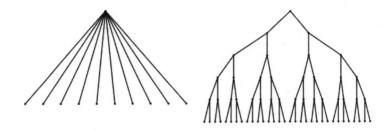

Two forms of structural virality

Source: Goel et al., 'The structural virality of online diffusion'

My own personal experience of going viral is somewhat limited, my most popular tweet being a picture of a quarter piece of cheese from Boxing Day 2018 that just happened to cost £π and weigh π-hundred grams. More than 2000 people 'liked' this post, which may say less about the hilarity of my visual humour and more about how bored

---

* I regret bringing up these two viral phenomena so early in the book, as any reader who has never heard of them is bound to go and look them up and be so gobsmacked they don't come back. Please do come back!

the general public are after Boxing Day dinner. Either way, I can't deny how seductive the thrill was of knowing that every five minutes when I checked my phone I'd have another glut of notifications. 'What shall I tweet about next?' I thought. What other hilarious visual trigonometry puns might I have up my sleeve? Alas, like catching lightning in a bottle, it seems the audience for cheese-based maths gags had peaked, and my next attempt at Twitter humour was greeted by my usual three likes and one retweet.

Kyle D Evans
@kyledevans

£2 per radian, solid pricing structure

THIS PIECE
£3.14
0.314kg

4:06 PM · Dec 26, 2018 · Twitter for iPhone

(I'm aware, by the way, that not every reader might understand the cheese joke, and also that the only thing less appealing than having a joke explained is having maths explained and a joke explained simultaneously. For this reason the explanation of the cheese gag can be found at the back of the book, as can the solutions to any maths puzzles in the book that are beyond GCSE level understanding, or are just too long or fiddly that they would disrupt the flow.)

If I really wanted to gain more traction on social media, one quick way might be to regularly drop emoji equation problems like the following:

(Source unknown)

You'll have seen this kind of thing before, undoubtedly accompanied by a glut of people fighting tooth and nail over the solution. I'm endlessly fascinated that we live in a world where millions of people wear their lack of mathematical ability as a perverse badge of honour, while the same people paradoxically delight in hurling abuse at any *idiot* who would have the temerity to disagree with what they think the answer is.

As a borderline millennial I've lived both with and without the internet, first having access to the 'worldwide web' at that transformative age of 16. The whole world of information at my fingertips! Oh the possibilities! Little did I know that in 20 years I'd be scrolling through endless examples of my auntie arguing about banana emojis with a 14-year-old Texan and calling me in for back-up.

In this book I'll present 55 notable cases of viral mathematics: from maths life hacks to Facebook fruit problems, fiendish exam questions to pre-internet playground classics. The solutions to most puzzles will directly follow the question, so to avoid spoilers you might want to put the book down at regular intervals while you work out the answers for yourself. There are some blank pages at the back to save you from scribbling on receipts, bank statements or your child's latest school report. There are also some maths tricks and 'life hacks' that I will explain the workings of; again, if you are desperate not to have the magic explained but to work it out yourself, please put the book to one side or skip forward to the next section. I won't be offended – I don't know you!*

If the emoji equations above elicited a queasy feeling in your stomach, fear not: I promise there will be many moments of sublime logical illumination to sweeten the pill. If you're a real puzzle enthusiast, or just extremely online, you'll have seen lots of these puzzles before, but I hope that even the most hardened puzzler will find a few new gems here. Very few of them come from my own imagination, and I've tried my hardest to credit the puzzle creator in all cases where they deserve it. In some cases, where the puzzle is completely derivative or has no redeeming features at all, I have made no such effort.

If, after all this, we're still no closer to discovering *why* some maths goes viral, at least we will have had fun trying.

Now, without further ado, let me lead the way. These maths tricks will *blow your mind!*

---

* Unless I do, in which case: hello!

THIS ONE COOL MATHS TRICK WILL BLOW YOUR MIND

Maths tricks and 'life hacks'

#1
Calculate 4% of 75.

Some of the most satisfying maths 'tricks' can be those that were under our noses for years and years but that we never noticed. Recently I was starting a lesson with an A-level maths class when one of my 17-year-old students entered, marched to the front of the room, picked up a whiteboard pen and proudly wrote in huge letters on the board:

4% of 75 is 75% of 4

The student then charismatically replaced the whiteboard pen lid (just about resisting the urge to drop the pen to the floor in the time-honoured 'mic drop' style) before turning to face the room to revel in the adulation of his peers. Well, half of his peers. Half of the class were as mind-blown as he was – and, perhaps, you were? – whereas the other half remained thoroughly nonplussed. They'd realized the truth of this little mathematical nugget years ago.

This, of course, is the trick mentioned in the introduction, and every few months something like this bubbles up on Facebook or Twitter, with varying levels of virality. Here is the example from 2019 that my student had seen, garnering tens of thousands of likes and retweets on Twitter:

Ben Stephens @stephens_ben · Mar 3, 2019
Fascinating little life hack, for doing percentages:

x% of y = y% of x

So, for example, if you needed to work out 4% of 75 in your head, just flip it and and do 75% of 4, which is easier.

○ 628   ↱ 10.7K   ♡ 23K   ⬆

Here are some suitably stunned replies:

This just broke my brain.

I used to teach maths for reporters and wish I'd had this explanation in my back pocket. I had other tricks for mathsphobes but this is far more elegant.

You have won Twitter today, Mr Stephens!

And here are some people who have no room for joy in their life:

Seriously? All you're saying is $3 \times 2 = 6$. As does 2×3. Do you really think people don't understand such a simple concept enough to know this? Good grief.

Why is this news? This is called 'propiedad conmutativa' (Commutative property) in spanish and it's taught in elementary school. Millennials discover... .

More about that '*propiedad conmutativa*' later. Let's break it down a little bit and first check that the trick definitely does work, before getting into why it works.

75% of 4

A fundamental of working with percentages is knowing their equivalent fractions and ratios: 75% is simply three-quarters, so we can very quickly see that three-quarters of 4 is **3**.

4% of 75

This is a little harder to calculate (without knowing the trick!), since 4% is not a 'nice easy' fraction. A method schoolchildren might be shown is to use 10% and 1% as starting points, using the fact that 10% of a number is ten times smaller than the original number, and 1% is ten times smaller again. In this case:

100%   of 75 = 75
10%   of 75 = 7.5
1%   of 75 = 0.75

To find 4% of 75 from here, we can put together four lots of 1%, i.e. 4 lots of 0.75 is **3**.

So if faced with the need to calculate 4% of 75, it's much easier to swap and calculate 75% of 4 instead. Don't fancy calculating 42% of 10? No problem – do 10% of 42 instead, much easier! It even works for percentages above 100. Try finding 25% of 400 – easy, right? A quarter of 400 is obviously 100. Our method says this should be the same as

400% of 25, and of course it is: 400% of 25 simply means four lots of 25: also 100. Calculations like this take us a little closer to the 'why', more of which shortly.

Try it yourself

(a) 73% of 10
(b) 12% of 25
(c) 16% of 75
(d) 44% of 5
(e) 13% of 25

I find this little mathematical trick very satisfying, not least because I still remember working it out for myself in maths class and peering around the class to see if anyone else had cottoned on. They hadn't! It seemed that only I knew this incredible secret – I'm not even sure if the teacher knew!

By the way, I don't claim to be some great genius in working the trick out for myself. Many a greater mathematician than I have gone much of their adult life without knowing this trick: in writing this book I surveyed my army of followers on social media to find out how many knew about this trick. Although they're probably slightly more maths-savvy than the average social media user, the results still showed nearly half who did *not* know that 4% of 75 was equal to 75% of 4:

Learnt it in school: 16.4%
Learnt it from social media: 25.2%
Learnt it from somewhere else: 15%
Did not know: 43.5%*

---

* You will have to ask Twitter why the percentages don't add up to 100%.

So if you had your mind blown by this, you're in good company. It was interesting that in the comments many teachers stated that they didn't know this until they began teaching, much as many people express dismay that they were never taught this in school. Now you know why: your teachers didn't know either! But that's fine; every day's a school day (especially if you work in a school). Let's zoom in a little closer to the detail of exactly *why* this works. In all of the above calculations I've shown how to find percentages of amounts using mental methods, as you would if you didn't have a calculator at hand.

Many countries – possibly most countries – implement systems where tax is added to the value of goods purchased at the point of payment. In England, where I live, this form of tax is usually applied to goods before the customer sees the price, but there are a few warehouse-style shops where VAT (value added tax) is not shown on shelf pricing, so that customers must work it out for themselves if they want to be prepared for how much they're really paying.

I still vividly remember accompanying my parents to such a shop as a child, where 'bulk buy' items were displayed without VAT, and being shown how to calculate the new price including VAT (which at the time was 17.5%) in my head. For example:

Price without VAT:	£40
10%:	£4 ('shift to the right', as we saw above)
5%:	£2 (halve it)
2.5%:	£1 (halve it again)
Price with VAT:	£40 + £4 + £2 + £1 = £47

However, in January 2011 the standard rate of VAT was changed to 20%, which it has remained since: a tragedy for maths teachers. Working out 17.5% was such a cool method: 10%, then halve it, then halve it again; in comparison, calculating 20% is a real bore. I don't know a lot about the economy, but I'd very much in favour of seeing the 17.5% rate come back, if only to make maths lessons (even) more interesting.

Mental arithmetic is a fantastic and under-appreciated skill, but *not* using a mental shortcut actually makes it easier to see under the bonnet here. One of the ways I found 4% of 75 earlier was to find 1% of 75, and then multiply this by 4. To put it another way:

$$\frac{75}{100} \times 4 = 3$$

Similarly, I could have found 75% of 4 by finding 1% of 4, and then multiplying that by 75. This is far from the best way to do it in your head, but mathematically it works just fine:

$$\frac{4}{100} \times 75 = 3$$

Now let's put those two calculations side by side:

$$\frac{75}{100} \times 4 = 3 \qquad \frac{4}{100} \times 75 = 3$$

Notice that both calculations involve multiplying 4 by 75, and dividing by 100. All that's different is the order in which the operations are carried out. Now, strictly speaking, division is the same thing as multiplication anyway: for example, multiplying by 2 is the same as dividing by a half.[*]

Since division is really multiplication in disguise, we can think of our 'trick' as two multiplications, rather than a multiplication and a division. All we need now consider is: does it matter what order we multiply in? Are three 2s the same as two 3s? Of course they are the same: this is a

---

[*] Think: how many 50p pieces are there in £5?

fundamental building block of mathematics known as *commutativity* (as seen in the smug tweet earlier). Note that not all mathematical operations have the commutative property: for example, 5 minus 2 is *not* the same as 2 minus 5, so subtraction is not commutative (nor is division).*

Since multiplication is commutative, 4% of 75 and 75% of 4 are simply asking you to carry out two multiplication operations in a different order, so by definition the outcomes will be the same.

If this trick was new to you before reading this chapter, it's probably because you're proficient in calculating percentage shortcuts in your head. It's only when you lay the calculations out in seemingly unnecessary detail that the 'magic' in the trick is revealed.

#2
Think of a whole number from 1 to 9.
Multiply it by 3.
Add 3.
Multiply by 3 again.
Add the two digits of your number together.
Finally, subtract 1. Whatever number you now have, turn to that page to continue the story...

---

* Addition is commutative, as Tom Lehrer put it in his song 'New Math':
And you know why four plus minus one
Plus ten is fourteen minus one?
Cause addition is commutative, right!

Hello! Congratulations, you correctly found your way to an answer of 8. This trick was cutely adapted in many different ways during the various Covid-19 lockdowns of 2020/21, including the following example which was tweeted by Devon & Cornwall Police when the first national lockdown was declared in the spring of 2020.

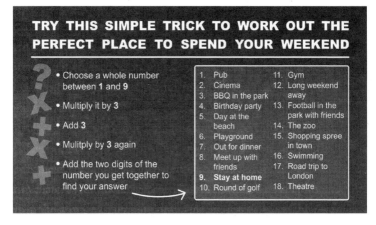

TRY THIS SIMPLE TRICK TO WORK OUT THE PERFECT PLACE TO SPEND YOUR WEEKEND

- Choose a whole number between **1** and **9**
- Multiply it by **3**
- Add **3**
- Mulitply by **3** again
- Add the two digits of the number you get together to find your answer

1. Pub
2. Cinema
3. BBQ in the park
4. Birthday party
5. Day at the beach
6. Playground
7. Out for dinner
8. Meet up with friends
9. **Stay at home**
10. Round of golf
11. Gym
12. Long weekend away
13. Football in the park with friends
14. The zoo
15. Shopping spree in town
16. Swimming
17. Road trip to London
18. Theatre

Credit: Twitter/DC_Police

The joke, of course, is that everyone's calculations will lead them inexorably towards number 9: stay at home, clearly the best advice in a global pandemic.

This trick is based on a feature of the 9 times table that you might remember from school, namely that the two digits of numbers in the 9 times table (up to 10 nines) always add up to 9.

$1 \times 9 = 9$

$2 \times 9 = 18$ $1 + 8 = 9$

$3 \times 9 = 27$ $2 + 7 = 9$

$4 \times 9 = 36$ $3 + 6 = 9$

$5 \times 9 = 45$ $4 + 5 = 9$

$6 \times 9 = 54$ $5 + 4 = 9$

$7 \times 9 = 63$ $6 + 3 = 9$

$8 \times 9 = 72$ $7 + 2 = 9$

$9 \times 9 = 81$ $8 + 1 = 9$

$10 \times 9 = 90$ $9 + 0 = 9$

If you need to add 9 to a number, you may have used the shortcut of adding 10 and subtracting 1 instead: it is very easy to add and subtract tens and ones, so the two operations of adding 10 and subtracting 1 are usually simpler than the single operation of adding 9. This process itself explains why the digits of two-digit numbers in the 9 times table add up to 9. The first number is 9, and after this you are simply adding 1 in the tens column and subtracting 1 in the units column. This leaves the sum of the two digits unchanged, i.e. still 9.

This very useful fact can be used in many ways, one of those being the clever way you might have been taught your 9 times table using your fingers. Firstly, hold up all ten fingers and number them from 1 to 10, left to right. Now, if you want to calculate, say, eight 9s, just lower the eighth finger and count how many fingers are raised on either side. In this case you'll see seven fingers raised to the left of the finger you've lowered, and two to the right. These numbers give you the tens and units parts of the answer you require: 72.

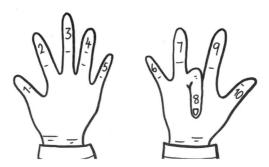

This is also the root of many maths magic tricks and 'mind reading' tricks such as the one above: instructions that give seemingly free choice but land the participant on a certain number. Let's have a look at why this particular set of operations in the 'Stay at home' graphic is bound to land on a multiple of 9. We could take every possible starting number and run it through the algorithm.

Disclaimer: Please try not to have a heart attack at the sight of the word 'algorithm' there. It's fair to say that between stories involving Facebook's dodgy dealings with Cambridge Analytica and the unfair grade distribution to some British schoolchildren in exams, algorithms have not had good press in recent years. But remember that an algorithm is simply a set of instructions – nothing more or less. Yes, Facebook does use many algorithms, not all of which have been entirely ethical. And the British government did use an algorithm to decide exam results for 16- and 18-year-olds who missed their 2020 exams due to the Covid-19 pandemic.* But an algorithm cannot be inherently evil – that allegation can only be levelled at the person or institution that created it.

---

* In truth, they tried to use an algorithm – which is perfectly reasonable – but overlooked the problem of the very limited historical data sets from which comparisons could be drawn for small cohorts, which essentially gave students from private schools an advantage over those from large schools and colleges. Eventually the whole thing was scrapped.

Where were we? Oh yes, the algorithm in the Devon & Cornwall Police tweet. Let's run every possible start number through it, using the magic of spreadsheets:

	×3	+3	×3
1	3	6	18
2	6	9	27
3	9	12	36
4	12	15	45
5	15	18	54
6	18	21	63
7	21	24	72
8	24	27	81
9	27	30	90

We can see from the right-hand column that, no matter where we start from, we end up at the familiar 9 times table. But it doesn't feel particularly elegant to draw up a table, does it? And we can't check that it *always* works by drawing a table either: even if we drew a table with 100 rows, the rule might break down on the 101st. Let's instead consider what happens to the general starting number.

The first step is to multiply by 3, so after this step we get to 'three lots of the number you first thought of'. Does that feel like a bit of a mouthful? Would it be OK if instead of 'the number you first thought of' and 'three lots of the number you first thought of' we simply use '*n*' and '3*n*'? If you answered 'yes', congratulations, you've agreed that algebra is useful! If you answered 'no' that's fine too, but, from this point on, whenever you see 'n' feel free to think in your head: 'the number I first thought of'.

With algebra, things look so much neater:

Starting number:	n
Multiply by 3:	$3n$
Add 3:	$3n + 3$
Multiply by 3:	$3(3n + 3)$, or $9n + 9$

Why is our end result, $9n + 9$, always a multiple of 9? Well, $9n$ is literally the algebraic definition of the 9 times table: 9 lots of n, where n is any number you like. Add 9 to a number in the 9 times table and you will simply land on the next number in the 9 times table.

Once you have seen the underlying algebra you could in fact generate your own set of instructions: as long as it ends up on $9n$ or $9n + 9$, you will have a set of two-digit numbers that can safely be added to guarantee a sum of 9 for the next part of the trick. Here's another example:

Think of a single-digit number:	n
Add 2:	$n + 2$
Double it:	$2n + 4$
Subtract 6:	$2n - 2$
Double it:	$4n - 4$
Subtract the number you first thought of:	$3n - 4$
Add 7:	$3n + 3$
Triple it:	$9n + 9$

This is probably less effective as an algorithm for a mind-reading trick, as it involves more steps and hence more scope for human error, but the effect is the same if the calculations are carried out correctly.

At this juncture it would be rude not to mention the most famous of all tricks based on the 9 times table.

#3

Think of a whole number from 1 to 9.

Add 2.

Multiply by 10.

Subtract 11.

Subtract the number you first thought of.

You now have a two-digit number.

Add the two digits, then subtract 5.

Whatever number you now have, assign it a letter in the following way: A = 1, B = 2, C = 3, etc.

Think of a *country* beginning with that letter.

Now move on to the next letter in the alphabet, and think of an *animal* beginning with that letter.

Finally, think of the colour of that animal, and turn over for a surprise...

There are no grey elephants in Denmark!

This is probably the best known and most oft-repeated example of a maths mind-reading trick. I have no idea where it originates from, but there's something appealingly memorable about the image of a Danish elephant that seems to have stuck this problem in the general human consciousness – I've witnessed my now-adult ex-students arguing about it on Facebook and I'm reliably informed by my younger, cooler friends* that it's been spotted on Instagram and TikTok too.

Curiously, the first hit I found when googling the phrase 'denmark elephant' was to the website playtivities.com, which claims that this trick will work '90% of the time'. Percentage figures that seem tenuous at best are quite a feature of viral social media maths, as we shall see throughout this book. I'm always annoyed when I see these percentages that seem quite clearly fabricated, so I've taken up a personal mission to investigate their correctness.

---

* OK, my nieces and nephews.

FACT CHECK!

There are of course two ways this trick can fail: either the person answering goes wrong with their maths, or they don't think of both Denmark and Elephant. Clearly this trick was first created to exploit the very high likelihood that a person will think of this combination of country and animal, but how many do? I surveyed a sample of 209 responders, made up of my students, colleagues and esteemed members of the fellowship of @kyledevans Twitter followers, and found the following results:

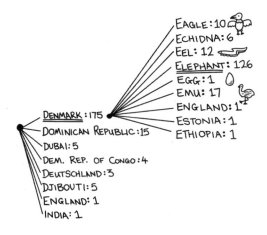

A few things to clear up first: clearly some people didn't read the questions properly (or they really think Estonia is an animal), and some people knew the trick and actively tried to say something else (one response read: 'Djibouti – DEFINITELY NOT DENMARK!')* Putting this aside,

---

* In recent years I've observed a high prevalence of the answer 'Djibouti', and I lay the responsibility for this widening of young people's geographical knowledge firmly with the popular and brilliant afternoon TV quiz show *Pointless*.

of the 209 people who responded, 175 said Denmark for the country, and, of those 175, 126 said Elephant for the animal. Bearing in mind that they have to say both Denmark and Elephant for the trick to come to a satisfactory conclusion, my evidence suggests that the trick will only fully work around 126 times out of 209, or 60% of the time: far less than the website predicts, even if everyone does their maths correctly.

Is a trick worthwhile if it works only 60% of the time? If Paul Daniels had accidentally hacked Debbie McGee into bloody pieces two times out of five, I doubt he would have become quite so much of a household name. Well, he would have, but for a different reason.

I was going to write here about the ubiquitousness of the Denmark–Elephant trick, and just how and why it is so well known around the world, so I asked my friend and experienced mathemagician Ben Sparks where he first heard about the Denmark–Elephant trick.

'The what?' he said.

'The Denmark–Elephant trick. You know, the one where you make someone think of an elephant from Denmark.'

'Oh, I know that trick, but I do it as orange kangaroo from Denmark.' Ha! Orange kangaroos from Denmark! What an idiot; nobody does it like that. Just to check, I googled the phrase 'denmark kangaroo orange' and… it turns out loads of people do it like that: there are just as many hits for 'denmark kangaroo orange' as there are for 'denmark grey elephant'. Isn't it interesting how we all see the world from our own entrenched perspective?

It turns out Ben leads his subjects towards a *European* country beginning with D, but then asks them to pick an animal starting with the last letter of their chosen country, then a colour starting with the last letter of their chosen animal. I didn't research the watertightness of the Denmark–Kangaroo–Orange path as I felt I'd exhausted the patience of my horde of Twitter disciples. Rest assured there's always bound to be someone out there who will think of kingfisher or killer whale though, especially once you've built up the attention of the whole family at the Christmas dinner table.

The Denmark–Elephant trick is a good example of the subset of maths magic tricks that are not guaranteed to actually work, but (arguably) work regularly enough for the reward to be worth the risk of it going wrong. Here's something similar – you'll need a calculator.

#4

On your calculator, multiply together ten random single-digit numbers. You can use the same number more than once.

You should now have a fairly long number.

Cover the first digit and add up all the other digits.

Now you either have a single-digit number or a two-digit number.

If you have a single-digit number, move to the next step. If you have a two-digit number, add those digits to make a single-digit number, then move to the next step.

Subtract your number from 9.

Now uncover the digit you've been covering. Ta-dah!

If that was hard to follow, here's an example:

Multiply ten single-digit numbers: $2 \times 3 \times 3 \times 5 \times 5 \times 5 \times 7 \times 7 \times 8 \times 8$
$= 7{,}056{,}000$
Cover the leading 7, and add all the other digits: $5 + 6 = 11$
This makes a two-digit number, so add the digits again: $1 + 1 = 2$
Subtract this from 9: $9 - 2 = 7$, which is the leading digit you're currently covering.

What's going on here then? How can the other digits in the number possibly have an effect on the first number? This all relies on the feature of the 9 times table that we discovered a few pages ago, namely that the sum of all the digits (or 'digital root') of a multiple of 9 is always 9:

$1 \times 9 = 9$

$2 \times 9 = 18$ $1 + 8 = 9$

$3 \times 9 = 27$ $2 + 7 = 9$

etc.

What we didn't consider earlier is what happens if we take the times table further:

$10 \times 9 = 90$ $9 + 0 = 9$

$11 \times 9 = 99$ $9 + 9 = 18$ $1 + 8 = 9$

$12 \times 9 = 108$ $1 + 0 + 8 = 9$

$13 \times 9 = 117$ $1 + 1 + 7 = 9$

etc.

It seems that multiples of 9 always have a digital root of 9, as long as when we are faced with a two-digit number or longer, we continue to add the digits (as seen for 99 above.)

Once we know this fact we can quickly decide if any number is in the 9 times table, in other words a multiple of 9, by investigating the digital root. For example, is 6,698,106 a multiple of 9? Let us calculate the sum of the digits:

$6 + 6 + 9 + 8 + 1 + 0 + 6 = 36$

36 is not a single-digit number, so we continue:

$3 + 6 = 9$

We arrive at a digital root of 9, meaning 6,698,106 is indeed a multiple of 9.[*] It is therefore also true that the digital root of the whole number with the first digit covered, plus the first digit, will similarly be 9:

(Cover the leading 6): $6 + 9 + 8 + 1 + 0 + 6 = 30$

30 has a digital root of 3, and $9 - 3 = 6$, i.e. the left-hand digit that we covered. We can now see how the trick works: by forcing you to generate a multiple of 9, I was able to safely predict that the digital root of your long number would be 9, and exploit this fact.

But… how did I force you to pick a long number that was a multiple of 9? Well, when you multiplied ten different numbers together, you probably hit a 9 somewhere along the way. But, even if you didn't, you probably hit two 3s, or two 6s (which contain a 'hidden' 3), or a 3 and a 6. Any string of numbers multiplied together with these ingredients will definitely give a result that's a multiple of 9.

Of course there's no guarantee. You may well choose to multiply together ten 2s, giving $2^{10} = 1024$, clearly not a multiple of 9. But the vast majority of people[†] will choose a number that is divisible by 9, leading to a successful and very satisfying outcome.

Speaking of satisfying outcomes…

---

[*] Note also that any number with a digital root of 3 is a multiple of 3.

[†] If people were to choose their ten digits truly randomly, this trick would have around a 93% chance of succeeding – calculations in the back of the book. My experience, though, is that it actually works more often, because participants don't choose their digits in a truly random way, they 'spread them fairly' across 1–9, making the trick much more likely to succeed.

#5

Calculate 42 × 21.

How would you go about carrying out this two-digit multiplication question? Depending on when you went to school, you might draw something like this:

```
    4  2
×   2  1
-------
    4  2
8  4  0
-------
8  8  2
```

Or this:

×	40	2
20	800	40
1	40	2

$800 + 40 + 40 + 2 = 882$

Or you may have noticed that $42 \times 21 = 2 \times 21 \times 21 = 2 \times 21^2$, and that 21^2 is $20^2 + 2(20) + 1 = 441$, so the answer must be 882. Just me?

However, in late 2020 a video that showed how to do this with 'Japanese multiplication' began to blow people's minds on TikTok:

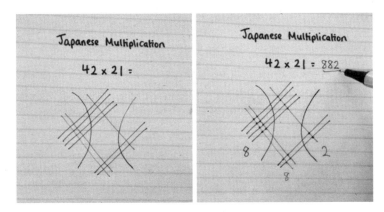

Credit: TikTok/jesslouisec

This method bubbles up fairly regularly on social media and it is quite wonderfully satisfying the first time you see it in action. The 42 is represented by the lines drawn in a north-easterly direction, and the 21 by the lines drawn in a north-westerly direction.

The number of intersections on the left represents the hundreds digit of your answer, the total number of intersections in the middle represents the tens digit, and the number of intersections on the right represents the units digit. This graphical representation of dry old multiplication can be quite joyous when you first encounter it.

But why weren't you shown this in school?! You may have realized that the example chosen for this particular viral video involves two numbers with relatively low-value digits in the tens and units places: 4, 2 and 1. This means we get a pleasingly low number of intersections: easy to count and no carrying required.

Here's a screenshot from my attempt at a viral 'Japanese multiplication' video that resolutely refused to take off. I wonder why?

79 x 86

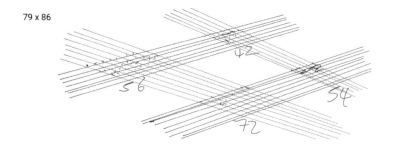

Here there are 54 intersections in the units position, and the 5 from this 54 needs to carry across to combine with the 114 intersections you've painstakingly counted in the tens position, giving 119. The 11 from this 119 carries into the hundreds position and combines with the 56 counted intersections to make 67. This 67 is followed by the 9 that was left behind from the 119 in the tens column, and the 4 from the 54 in the units column, to give 6794. Not quite so satisfying, is it?

Clearly using this method for this type of example creates a huge headache: when my wife found this calculation in my notebook she scolded our six-year-old for mindlessly scribbling on Daddy's work. It's also incredibly arduous to count up the number of intersections when 7 lines cross 8 lines – it's almost as if there might be an easier way to do this… multiplying 7 by 8 perhaps? In fact, the calculations required when carrying out 79 × 86 with 'Japanese multiplication' are 7×8, 7×6, 9×8 and 9×6, plus some carrying and multiplying by 10. Perhaps we might replace all those lines with something more compact that does the same job?

×	80	6
70	5600	420
9	720	54

5600 + 420 + 720 + 54 = 6794

I am happy to concede that the 'Japanese˙ multiplication' method is very pleasing when all the digits involved are small. Or, in other words, it adds a visually pleasing aspect to calculations that would be relatively uncomplicated anyway. Somehow I don't think that pitch is going to make me the next viral sensation though…

---

*   By the way, I can find no evidence of this method originating from Japan, nor any evidence of it being taught any more prevalently in Japan than in any other country. If we're going to give it a made-up name, I'm sure we could do a lot better. Klingon multiplication? Tumultiplication? Multiplicationy Mcmultiplicationface?

2

IT WAS DIFFERENT IN MY DAY

Pre-internet viral maths

In these days of our complete immersion in online tips, tricks and life hacks, it's easy to think that we never would have seen such things without constant access to the information superhighway. But anyone who was educated pre-smartphones* will tell you that riddles, puzzles and tricks were rife in playgrounds and workplaces long before those tiny boxes of plastic and metals started stealing our attention.

I'm always surprised and saddened, on the odd occasion I walk past my local secondary school at home time, to see how many teenagers are instantly drawn to their smartphones upon leaving school. It wasn't like that in my day (cue nostalgic music from the Hovis advert…) – we had to learn how to talk to each other on the walk home. Admittedly 90% of that conversation was football, who fancies whom and last week's episode of *The Fast Show* repeated sketch by sketch. But the remaining 10% would very often be riddles, puzzles and brain-teasers.

Often these were short Hobbit-esque language-based riddles: *What gets wetter the more it dries? What can you hold in your left hand but not*

---

* The first time I ever used the internet was when I signed up for 'internet club' which ran in the school library at lunchtimes, in which you'd be allotted five minutes to type a 50-digit long URL from the back of your journal into AOL.com and wait for the steam-powered dial-up to find your chosen website. Yes, I'm 85 years old.

your right hand? Sometimes they were more elaborate language-based conundrums: *You're in a locked prison cell with no windows and only a single wooden chair: how do you escape?*[†]

My friends and I also played a game that involved constructing elaborate murder scenes for each other, to which we would 'play detective' and try to surmise from the scene what happened: *A man is found dead in a phone box with his bloodied arms thrust through the glass panels on both sides. There's a fishing rod in the corner of the phone box and a woman's voice on the hanging phone saying 'hello? hello?' WHAT HAPPENED?*[‡] I once caught some teenage students playing this game and attempted to pose them the above example, but by the time I'd spent five minutes explaining the concept of a phone box they'd lost interest and moved on to something else.

In amongst all of this were the inevitable number-based riddles: *As I was going to St Ives I met a man with seven wives: each wife had seven sacks; each sack had seven cats; each cat had seven kits. Kits, cats, sacks and wives, how many travelling to St Ives?*[§]

I was never particularly fond of the St Ives riddle, for the following two reasons above all others. Firstly, the 'ahhhh, caught you out!' nature of the solution doesn't really work, as it's not made very clear that the greedy man and all his wives are travelling in the opposite direction to you. It's perfectly possible to meet someone when they're

---

* A towel and your right hand, respectively.

† Rub your hands until they're sore ('saw'), use the saw to cut the chair in half, two halves make a whole ('hole'), climb out of the hole to safety. Some people then add shouting until your voice is hoarse ('horse') and riding off to freedom, but that's just ridiculous.

‡ The man had been fishing and excitedly phoned his wife afterwards: 'I caught one THIS big!'

§ The answer is 1 person: the narrator of the riddle. Everyone else is travelling *away from* St Ives.

walking in the same direction as you are! Especially if there's something really exciting going on in St Ives (unlikely, I admit). Secondly, in this situation we're led to believe that each wife is carrying 7 sacks, each of which contains 7 cats and 49 kittens. An average cat weighs about 4kg and a kitten around 0.5kg; this means each woman is carrying around approximately 50kg in each sack, around 350kg in total, or roughly the weight of a large grizzly bear. I certainly have some questions. Do I think about these things too much?

Leaving Cornish polygamy and animal cruelty aside, a number-based problem that sticks in my memory as a cause of particular frustration is the following, usually known as the 'missing dollar' or 'missing pound' riddle.

#6

Three women meet and go out for lunch. At the restaurant they each order a meal for £10 and place three £10 notes on the table when it's time to pay. The waiter takes their money to the till, where the manager informs him that there is in fact a special offer on: 3 meals for £25. So the waiter takes five £1 coins back to their table as change. Each woman decides to take a pound back each, leaving £2 as a tip for the waiter.

Now, the women came with ten pounds each: £30. They paid nine pounds each for their meals, totalling £27, plus another £2 for the waiter, making £29 altogether. Where did the other pound go?

This famous riddle appears to be undestroyable; I remember first hearing it in the playground in the late 1990s, and promptly sharing this new and exciting brain-teaser with everyone I met. In fact, my mate Phil was so impressed by the disappearing-pound riddle that he wanted to set up a scheme where we'd drive around the country, exploiting this loophole to embezzle pound coins from unwitting businesses and riding off into the sunset. Phil of course overlooked one or two glaring problems:

- There were only two of us, not three.
- We were 14 and couldn't drive.
- You can only pull off the scheme at restaurants that offer three meals for £25.
- Even when you have found the restaurant, you then have to spend £25 (or £27?) just to create the missing pound.
- You don't actually gain the missing pound anyway, since by definition it is missing.

Other than those problems it was a foolproof plan. Little did I know at the time that this riddle was already more than 60 years old, dating back to the 1930s at least. In the intervening years I've seen it appear in British adult comic *Viz** (see page 28) and it recently went (inevitably) viral on our old friend the video-sharing social network TikTok.

---

* My all-time favourite piece of mathematical content in *Viz* was a humorous story about the six times table plummeting in value and British parliament having to instigate a short-term emergency table in its place, whereby $1 \times 6 = 2$, $2 \times 6 = 4$, etc. This came with an inevitable 'winners and losers' side-panel, where winners featured domino manufacturers who would save ⅔ on ink, and losers included cricketers whose boundary shots would now only be worth 2 runs.

At the time of writing, googling the phrase 'missing pound riddle' returns news stories from the London *Metro* in 2018, *The Sun* in 2016, *Cambridge News* in 2020 and the *Daily Star* in 2015, on the first search page alone. If you're reading this book in 2025 (are there flying cars yet?) I would not be surprised if searching for the phrase 'missing pound riddle' returns results from 2021, 2022, 2023 and 2024 – although those newspapers I mentioned above probably won't exist.

> Over the years I drifted apart from my friend Phil and his cunning plan, but I think he'd have been proud of how I managed to fleece a fiver from a university friend, years later.
>
> We were at the student union and noticed tickets were on sale for a band we wanted to see. My mate didn't have cash on him, so I said he could buy the drinks on the night. As it happened we enjoyed the bands too much to spend long at the bar and just had a couple of drinks each.
>
> On the walk home, my mate – not a maths undergrad! – asked how much he owed me. I explained that I'd bought two tickets at £10 each; he'd bought four pints at £2.50 each. So I'd bought two tickets, and he'd spent a tenner – the equivalent of one ticket. Since I'd bought two tickets and he'd bought one, he owed me for one ticket. And he gladly handed over a £10 note. (Don't worry, I explained it the next day and gave him £5 back.)

The missing-pound riddle is in many ways the perfect piece of viral maths, in that it falls into the category of problem that I have snappily christened PWSLSTS: problems where the solution is less satisfying than the set-up. The brilliant sci-fi novel *The Prestige*, by Christopher Priest – later turned into one of Christopher Nolan's finest films – concerns two rival magicians desperate to discover the secret to the final stage, or 'Prestige', of each other's disappearance tricks. The agony of not knowing, which we've surely all experienced when being shown a brilliant new magic trick, drives the two illusionists to extreme lengths.

But discovering the secret is rarely as satisfying as not knowing, as the missing-pound riddle surely demonstrates.

Perhaps this is an area where maths tricks and magic tricks diverge, because in a magic trick it's not impressive to see a dove disappear, or even a person disappear: you need to see it come back. The 'Prestige' in the missing-pound riddle, like the percentages trick that opened Chapter 1, is actually pretty disappointing, certainly in comparison to the wonder that the puzzle itself sets up. But that doesn't matter: the telling of the riddle, and watching someone squirm while they try to puzzle it out, is the real enjoyment.

There is, of course, **no missing pound**. The women jointly paid £27 for a £25 meal, and their £2 overpayment is the waiter's tip. That's it. Adding the £2 change to the £27 price of the meal takes you tantalizingly close to £30, but it's meaningless to do so. The mention of £30 has no need to be in the riddle at all: it is planted at the start purely as a red herring.

Slightly disappointing, right? Just like any magic trick, seeing what's going on behind the scenes is rarely as thrilling as the exquisite agony of squirming while you try to work it out for yourself. This is a common trait of viral maths tricks and puzzles, whether pre- or post-internet: if a problem is intriguing enough in its set-up, it doesn't really matter how satisfying the resolution is; it has probably already gone viral by that stage.

I believe the missing-pound problem first spread amongst my friendship group when our maths teacher Ms Henderson told it to one of us in class. I remember she took great delight in trying to wind up and frustrate us little smart-arses, and she's been a great role model for my own teaching career. The next problem, however, I did not encounter until my teacher training, though many readers will recognize it from their own schooldays. You'll need a pen and paper.

#7

Take any three-digit number, as long as it isn't palindromic (the same forwards as backwards).

Turn the number backwards to make another three-digit number, and subtract the smaller number from the larger.

Take your new answer (call it 'A') and turn it backwards, to make another 3-digit number (careful: if your number appears to be two-digits, i.e. 99, think of it as 099 and turn it backwards to give 990).

Finally, add this number to A to give your *final answer*.

Square root your answer, then subtract 1 and turn to that page for a surprise.

If you don't know how to do a square root, divide your answer by 33 and then subtract 1.

Well done! You correctly carried out the 1089 trick! Amazingly, if you carried out the steps correctly, every starting point will lead to the answer of **1089**.

This is one of the best-known tricks in all of mathematics (or mathemagic?) and it even inspired the title of David Acheson's brilliant book *1089 and All That*. Acheson writes of first discovering the problem in a magazine in 1956, at the age of ten, and I have to say that when I discovered it as a trainee teacher fully 50 years later I was no less impressed. Of course, because I was on a teacher training course, the next challenge was to prove *why* it works (not that I wouldn't have wanted to do so anyway).

The first step is to work out how to express a general three-digit number algebraically. Our first instinct might be to write abc, but in algebra this means *a multiplied by b multiplied by c*, which is not what we mean. What we really need is $100a + 10b + c$, where a, b, c represent the digits in the hundreds, tens and units places, respectively. Then we can turn the number round to give $100c + 10b + a$, and subtract:

$$
\begin{array}{rllll}
 & 100a & + & 10b & + & c \\
- & 100c & + & 10b & + & a \\
\hline
 & 100a - 100c & & & + & c - a
\end{array}
$$

This expression can be rearranged and simplified to give $99a - 99c$, which we might prefer to write as $99(a - c)$. Note that $a - c$ is the difference between the first and last digits of the original number you chose, which is at least 1 (because I didn't allow you to choose a palindromic number) but at most 9 (the largest possible difference between two single-digit numbers). Therefore the expression $99(a - c)$

represents the first nine multiples of 99, and these are the only values you can be on after the first part of the trick:

$1 \times 99 = 099$
$2 \times 99 = 198$
$3 \times 99 = 297$
$4 \times 99 = 396$

Before I go any further, what do you notice about the values in the 99 times table?

$5 \times 99 = 495$
$6 \times 99 = 594$
$7 \times 99 = 693$
$8 \times 99 = 792$
$9 \times 99 = 891$

Repeatedly adding 99 has the effect of adding 1 in the hundreds column but subtracting 1 in the units column. This means that, as we move through the 99 times table, we are adding 1 in the left-hand column but subtracting 1 in the right-hand column. Not only that, but the first and last digit of each of these numbers adds up to 9. This means that when we reverse any of these numbers and add, we get something like:

```
   0  9  9        2  9  7        6  9  3
+  9  9  0     +  7  9  2     +  3  9  6
_____    _____    _____
```

In the units column we always get a total of 9, and in the hundreds column we also always get 9. In the tens column we always get two 9s. So the sum will always be 9 hundreds, plus 18 tens, plus 9 units. In other words, 1089.

I've seen the 1089 trick in many and various places over the years, including more than once on British TV (*Mr Drew's School for Boys, QI*) and I honestly never get tired of it. To quote Samuel Johnson, I believe: 'When a man is tired of the 1089 trick, he is tired of life.' Like any magic trick, it can be dressed up in various ways to add more excitement, but at its core there is a thrill that I'm sure will continue to deliver for years to come. Certainly I believe it to be a borderline criminal offence to teach students three-digit addition and subtraction without bringing in a bit of 1089 magic somewhere along the line.

FOUR MORE FROM THE PLAYGROUND
(Answers at the back of the book)

1. A clock takes 2 seconds to strike 2 o'clock. How long does it take to strike 3 o'clock?

2. You live in a special house with four south-facing walls. One day, when leaving your house, you see a bear. What colour is it?

3. You are a bus driver. You leave the depot with four passengers on board. You then drive to Aylesbury where three people get on and two get off. Finally, you drive to Northampton where two get off and five get on. What's the bus driver's name?

4. Do these simple sums in your head: Double 2. Double 4. Double 8. Double 16. Double 2. Double 4. Double 8. Double 16. Double 2. Double 4. Double 8. Double 16. **Now shout out the name of a vegetable!**

The third of those playground teasers above appears to be riffing on the acknowledged puzzle template of passengers getting on or off a bus or train. These questions seem to hold a lasting mathematical appeal, as we can see from the following recent viral problem.

#8

There were some people on a train. 19 people get off the train at the first stop. 17 people get on the train. Now there are 63 people on the train. How many people were on the train to begin with?

This English primary school question bounced around the internet in 2017, harvesting an alarmingly large number of comments on the Facebook group 'Parents Against Primary Testing' before being reported on by Huffington Post and various other news outlets.

I did promise myself when writing this book that I would never use the phrase 'I can't see how anyone can argue with this', especially since our collective experience of social media is that people will argue that black is white.* But with this one, I'm sorry, but I don't understand how anyone can see an answer that isn't **65**. Regardless of how you feel about testing in primary schools, this question seems to be asking students to appreciate that 19 people getting off the train while 17 get on represents a net loss of 2. At the end there are 63 on the train, so there must have been 65 to begin with.

Whilst scrolling through the original Facebook post (time I will never get back) I found a large number of people thought the answer was 46, and it took me a long time to understand where the 46 comes from since hardly any of them were willing to show their workings.

---

* Ah, to think that only a few years ago I used to use 'arguing that the world is flat' as a hypothetical illustration of people who will argue about anything. Now flat-earthers are only about the world's 3rd or 4th most ridiculous conspiracy theorists.

Where's the fun in that? Much more important to just have a Facebook slanging match with a stranger.

I think I've cracked it now though: if you *don't* couple the −19 and +17 together to make a net loss of 2, then one step before the end of the puzzle the train has 46 people on it (63 − 17). You would then need to construe 'How many people were on the train to begin with?' as referring to that particular moment before the 17 people get on, but after the 19 people had got off. I still think this is a real stretch, but I do at least now feel I've walked a mile in someone else's shoes and seen the world from their perspective, which of course is always important. Unless they're a flat-earther.

By the way, I concede that this is a hard question for six- and seven-year-olds, who it seems to have originally been aimed at. But what's wrong with hard questions? Yes this question takes a long time if you don't spot the net loss of 2, but it's by struggling with a difficult question that we become illuminated on how to do better in the future. We'll explore more viral exam questions later in the book.

Before the internet, when the very nature of 'virality' was different, there were all sorts of weird and wonderful ways that recreational maths could enter our lives. I remember a friend returning to school the first Monday after a Christmas break and producing six crumpled cards from their coat pocket, eager to read all of our minds.

#9

Pick any number on any of the cards, but don't say it out loud.

1	3	5	7	9	11	13	15
17	19	21	23	25	27	29	31
33	35	37	39	41	43	45	47
49	51	53	55	57	59	61	63

4	5	6	7	12	13	14	15
20	21	22	23	28	29	30	31
36	37	38	39	44	45	46	47
52	53	54	55	60	61	62	63

16	17	18	19	20	21	22	23
24	25	26	27	28	29	30	31
48	49	50	51	52	53	54	55
56	57	58	59	60	61	62	63

2	3	6	7	10	11	14	15
18	19	22	23	26	27	30	31
34	35	38	39	42	43	46	47
50	51	54	55	58	59	62	63

8	9	10	11	12	13	14	15
24	25	26	27	28	29	30	31
40	41	42	43	44	45	46	47
56	57	58	59	60	61	62	63

32	33	34	35	36	37	38	39
40	41	42	43	44	45	46	47
48	49	50	51	52	53	54	55
56	57	58	59	60	61	62	63

Now keep all of the cards that feature your number, and discard all of the cards that don't.

Add together all of the numbers in the top left of the cards you kept.

This is usually known as the 'mystery calculator' and is a staple of British Christmas cracker toys (and certainly a more satisfying distraction than the plastic comb, leaping frog or fortune-telling fish). When performing the trick live, the magician of course adds the numbers in the top left corners before proudly declaring the player's number out loud.

This left me absolutely baffled as a child, and in the years before the internet I was left stumped for years; no one at home would have had the time or inclination to help me work it out, and honestly I never thought of asking a maths teacher. There's a meme-like story that occasionally does the rounds on the internet about a child whose parents repeatedly used the apparently wise aphorism: 'Knowledge is power; France is bacon.' The child understood the supposedly profound nature of the first half of the phrase, but could never understand why France was bacon, and felt silly asking in front of people who seemed to sagely nod in agreement whenever it was said out loud. It was only years later, as an adult, that they first saw the phrase written down: 'Knowledge is power' – Francis Bacon. Anyone who lived through the pre-internet years will remember similar cases of not knowing something and *just having to wait*.

So what's the trick to the mystery calculator? First it's worth noticing that the numbers on the cards are not random; they do follow a pattern. Card 1 (that is, the card with '1' in the top left corner) contains all the odd numbers, Card 2 has numbers in clumps of 2 with gaps of 2, Card 4 has numbers in clumps of 4 with gaps of 4, and so on.

The genius step is realizing that any whole number can be written in one unique way as a combination of powers of 2 (which are the 'doubling' numbers: 1, 2, 4, 8, etc.). For example:

$7 = 4 + 2 + 1$
$14 = 8 + 4 + 2$
$17 = 16 + 1$
$31 = 16 + 8 + 4 + 2 + 1$

You'll find that the number '7' appears on cards 4, 2 and 1, so that when the player chooses these cards the magician can simply add together the numbers in the top left corner. Look again at the cards above. 63 appears on every card, because 63 is the sum of 32, 16, 8, 4, 2 and 1 (any number that is 1 less than a power of 2, such as 63, 31 or 7, will always be the sum of all the previous powers of 2).

My next thought was what an amazing coincidence it was that all of the required values fit snugly on to six cards of 32 values on each. To try to explain this trick to someone many years later, I made a much simpler version of the trick with just three cards:

Card 1:   1, 3, 5, 7
Card 2:   2, 3, 6, 7
Card 4:   4, 5, 6, 7

Writing each number in terms of powers of 2, we have:

$1 = 1$

$2 = 2$

$3 = 2 + 1$

$4 = 4$

$5 = 4 + 1$

$6 = 4 + 2$

$7 = 4 + 2 + 1$

When we look at the composition of all the numbers from 1 to 7 as powers of 2, we see that 1, 2 and 4 all appear four times each, so that cards 1, 2 and 4 will all have four values each on them. This simpler three-card version makes it easier to see the link between the top left number on each card and the other numbers on the card.

The classic 'mystery calculator' consists of six cards of 32 values each, but simpler versions could be made with five cards of 16, four cards of 8, three cards of 4 (my simpler version above) etc. You could even have a version with seven cards of 64 values, eight cards of 128, and so on. Six cards of 32 seems to be the most satisfying arrangement, so that the cards don't look too complicated but the trick doesn't seem overly simple either.

Maths and magic have often been easy bedfellows, including in the following satisfying trick which readers of a certain age may remember from a 1990s David Copperfield routine:

#10

Choose a number from the clock face, i.e. 1–12.

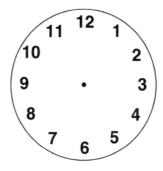

Spell your number and count the number of letters in the spelling. Whatever number you counted to, put your finger on that number on the clock.

Next, spell the number you're now on, and count onwards by that number. For example: if you were on 3, this has five letters so count on by 5. 3 + 5 = 8.

Whatever number you are now on, carry out the process once more: spell the number, count the letters, and move forward by that number.

You're now pointing to the number on the clock at the front of the book, on the contents page.

I've always loved this trick, and variants of it. You can probably see after thinking about it for a little while that, at every iteration of the trick, numbers on the clock face converge. At the first stage there are 12 numbers you can choose from, but all the numbers on the clock face when spelt out in English contain either three, four, five or six letters:

	Letters
One	3
Two	3
Three	5
Four	4
Five	4
Six	3
Seven	5
Eight	5
Nine	4
Ten	3
Eleven	6
Twelve	6

This means that a player's starting position on the clock face can only be 3, 4, 5 or 6. We can expand our table to show all the possible positions that a player can find themselves in throughout the trick:

	Starting position	After 1st move	After 2nd move
One (+3)	3	3 + 5 = 8	8 + 5 = 1
Two (+3)	3	3 + 5 = 8	8 + 5 = 1
Three (+5)	5	5 + 4 = 9	9 + 4 = 1
Four (+4)	4	4 + 4 = 8	8 + 5 = 1
Five (+4)	4	4 + 4 = 8	8 + 5 = 1
Six (+3)	3	3 + 5 = 8	8 + 5 = 1
Seven (+5)	5	5 + 4 = 9	9 + 4 = 1
Eight (+5)	5	5 + 4 = 9	9 + 4 = 1
Nine (+4)	4	4 + 4 = 8	8 + 5 = 1
Ten (+3)	3	3 + 5 = 8	8 + 5 = 1
Eleven (+6)	6	6 + 3 = 9	9 + 4 = 1
Twelve (+6)	6	6 + 3 = 9	9 + 4 = 1

You can see that after the first move a player can only find themselves at 8 or 9, and both of these converge on 1 after the next move (because 8 is 1 less than 9, but 'eight' has one more letter than 'nine'.)

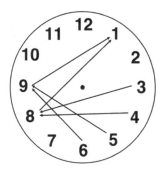

I call this convergence on one eventual end point the 'grey Play-Doh' phenomenon. Anyone whose children have ever played with Play-Doh will know that, from the moment the pristine primary-coloured pots of clay are purchased, they are on an inexorable slide towards coalescing into one giant grey lump. The separate colours can only converge; you can't pull them apart again. The toothpaste is easy to get out of the tube but impossible to get back in again. Am I mixing too many metaphors?

To see the grey Play-Doh phenomenon in action in a different context, please observe the following table, which I generated by shuffling my son's Uno cards (with the 0 cards and wild cards removed) and then dealing them into a grid. Your challenge is to pick a card in the top row, move forward by the number of squares shown on the card, and repeat this process until you can't move on any further. (When you get to the end of a row, start the next row on the left-hand side, like reading a book.) So if you were to start with the 8 in the middle of the top row, it would take you to the 9 in the second row. Now carry on in the same way until you cannot move further.

4	7	5	1	7	9	8	2	2	1	1	8
7	8	9	2	1	1	4	6	6	5	6	7
4	5	3	8	6	9	<u>8</u>	3	1	2	1	6
4	1	9	9	6	7	9	7	4	2	2	3
5	4	8	5	4	6	9	3	3	3	8	5
8	6	9	2	4	5	2	5	3	3	7	<u>7</u>

You're at the underlined 7, right? This card trick is known as a *Kruskal count*, named after its inventor, the mathematician Martin Kruskal. Like so many maths games, tricks and puzzles, it was popularized by Martin Gardner in the magazine *Scientific American.** Look at the 2s and 1s in the top right corner. Notice that all of these potential start points run very quickly into the 8 in the very top right: immediately five of the possible 12 start points have coalesced. The second 7 in the top row also runs directly into that same 8, so that's half of the possible start points joining up before we've even left the top row. Actually, this 7 has a 1 to its left, and the 4 in the very top left also runs into the same 7, so that's most of the top row accounted for.

In fact, the 7 and 5 in the top left also quickly join with the above path, as they both run into the patch of 2s and 1s in the top right. All of this only leaves two starting values that escape the top row without joining the same path – those are the 9 and 8 right in the middle of the row. Since the 9 is one place behind the 8, they quickly join forces in a

---

* There's also an excellent paper by James Grime on some of the maths behind the Kruskal count; find it at www.singingbanana.com/Kruskal.pdf

new path and move to the 9 in the second row, but by the underlined 8 in the third row, even this one rogue path has joined with the others. After this, it's an inevitable slide towards the final 7.

The Kruskal count has advantages and disadvantages when considered as a magic trick. On the one hand, it falls into the Denmark–Elephant category of tricks that aren't guaranteed to work: it is possible to deal out a grid where different starting points will remain on different paths all the way to the bottom row.* On the other hand, it can be done spontaneously with any pack of cards and no preparation required, which looks really impressive. Additionally, if a participant makes a mistake near the top of the grid and miscounts (hi Mum and Dad!), due to the grey Play-Doh phenomenon there's a good chance that the new path they erroneously find themselves on will still meet up with the most common path.

The Copperfield clock trick is a twist on the Kruskal count with an extra level of challenge introduced in that participants move around a closed cycle, drastically changing the 'inevitability' of paths joining.

After performing variations of this trick at my own live shows, it eventually occurred to me that I was only seeing this trick from one perspective: that of an English speaker. It didn't take too many performances before I encountered a volunteer who wanted to spell the letters out in their favoured tongue, namely Spanish:

---

* It's incredibly unlikely that this would happen with my Uno deck of 72 cards, as a higher number of cards makes it increasingly likely that paths will join together. The trick is usually done with a regular deck of 52 playing cards, where the chances of failure are higher, though still slim.

	Starting position	1st move	2nd move	3rd move	4th move
Uno (+3)	3	3 + 4 = 7	7 + 5 = 12	12 + 4 = 4	4 + 6 = 10
Dos (+3)	3	3 + 4 = 7	7 + 5 = 12	12 + 4 = 4	4 + 6 = 10
Tres (+4)	4	4 + 6 = 10	10 + 4 = 2	2 + 3 = 5	5 + 5 = 10
Cuatro (+6)	6	6 + 4 = 10	10 + 4 = 2	2 + 3 = 5	5 + 5 = 10
Cinco (+5)	5	5 + 5 = 10	10 + 4 = 2	2 + 3 = 5	5 + 5 = 10
Seis (+4)	4	4 + 6 = 10	10 + 4 = 2	2 + 3 = 5	5 + 5 = 10
Siete (+5)	5	5 + 5 = 10	10 + 4 = 2	2 + 3 = 5	5 + 5 = 10
Ocho (+4)	4	4 + 6 = 10	10 + 4 = 2	2 + 3 = 5	5 + 5 = 10
Nueve (+5)	5	5 + 5 = 10	10 + 4 = 2	2 + 3 = 5	5 + 5 = 10
Diez (+4)	4	4 + 6 = 10	10 + 4 = 2	2 + 3 = 5	5 + 5 = 10
Once (+4)	4	4 + 6 = 10	10 + 4 = 2	2 + 3 = 5	5 + 5 = 10
Doce (+4)	4	4 + 6 = 10	10 + 4 = 2	2 + 3 = 5	5 + 5 = 10

In Spanish we do reach the grey Play-Doh point ('*el Play-Doh gris*?') but it takes nearly two full trips around the clock face to get there. After just one move, all the potential start points have congregated at 7 and 10, but it takes another whole trip around the clock for the two streams to alight on 10.

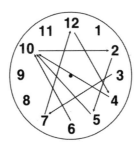

Now let's try that again, but in French this time:

	Starting position	**1st move**	**2nd move**	**3rd move**	**4th move**
Un (+2)	2	**2 + 4 = 6**	**6 + 3 = 9**	**9 + 4 = 1**	**1 + 2 = 3**
Deux (+4)	4	4 + 6 = 10	10 + 3 = 1	1 + 2 = 3	3 + 5 = 8
Trois (+5)	5	5 + 4 = 9	9 + 4 = 1	1 + 2 = 3	3 + 5 = 8
Quatre (+6)	6	**6 + 3 = 9**	**9 + 4 = 1**	**1 + 2 = 3**	3 + 5 = 8
Cinq (+4)	4	4 + 6 = 10	10 + 3 = 1	1 + 2 = 3	3 + 5 = 8
Six (+3)	3	3 + 5 = 8	8 + 4 = 12	12 + 5 = 5	5 + 4 = 9
Sept (+4)	4	4 + 6 = 10	10 + 3 = 1	1 + 2 = 3	3 + 5 = 8
Huit (+4)	4	4 + 6 = 10	10 + 3 = 1	1 + 2 = 3	3 + 5 = 8
Neuf (+4)	4	4 + 6 = 10	10 + 3 = 1	1 + 2 = 3	3 + 5 = 8
Dix (+3)	3	3 + 5 = 8	8 + 4 = 12	12 + 5 = 5	5 + 4 = 9
Onze (+4)	4	4 + 6 = 10	10 + 3 = 1	1 + 2 = 3	3 + 5 = 8
Douze (+5)	5	5 + 4 = 9	9 + 4 = 1	1 + 2 = 3	3 + 5 = 8

By the second move we seem to have three distinct 'streams': all numbers have settled upon 1, 9 or 12, and we would hope for these streams to join further as we continue the process, eventually coming together in one unifying stream as we saw in the English and Spanish versions. But alas! It will never happen. Look at the entries in bold. Imagine if you and your *ami* both played the game at the same time, with you thinking of the number 'un' and your friend thinking of 'quatre'. You put your finger on '2', because 'un' has two letters, and then count on four places, because you're now on 'deux', and 'deux' has four letters. This lands you on '6'. But your friend started with their finger on '6', because they were thinking of 'quatre', and 'quatre' has six letters. You take one move just to get to your friend's starting position, by which point they have moved on. You will eternally find yourself one move behind your friend, endlessly chasing them around the clock face in a futile dance. The trick fails in French!

The trick nearly works in German, with all starting numbers eventually falling into a satisfying loop of 1 → 5 → 9 (*eins* → *fünf* →*neun*), except for if you started from seven, in which case the loop you end up in is 2 → 6 → 11 (*zwei* → *sechs* → *elf*). Not so lucky after all. Turkish crashes almost immediately due to the very short names of some numbers: starting from two, or *iki*, leads you to three, or *üç*. These combined moves take you to five, but five is a possible starting point, so we will have two chains out of sync, like in the French example above. Vietnamese and Tagalog appear to fail for similar reasons, with the extra sticking point that twelve in Tagalog is *labingdalawa*, literally 'ten-two', which has twelve letters. Anyone choosing to start from the top of the clock will be stuck there eternally.

Try it yourself

I'd love it if readers who can speak other languages would try the trick and let me know if it works. The language will need to use the Latin alphabet (A–Z), or at least have a phonetic version that uses the Latin alphabet.

The idea of giving a volunteer a seemingly 'free' choice, then guiding them towards a satisfying surprise result is known as 'equivocation' or sometimes 'magician's choice'. We've already seen it in the 9 times table tricks in Chapter 1, and the 1089 trick in this chapter. Whether magical maths or just plain old magic, there's a timeless charm to this method: even when you know it's happening to you, it's oh so satisfying to be led along.

3

BACK TO SCHOOL

Viral exam questions and classroom conundrums

Do you remember the days of the old schoolyard? We used to laugh a lot. And then, five or so years later, we'd sit a series of written papers to test how well we remembered it all.

I took a borderline masochistic pleasure in exams, albeit more in the drama of the preparation than sitting the actual paper. I would train like a prize fighter coming up to a life-changing bout, practising every single past paper question I could get my hands on while the music from the *Rocky* training montages played in the back of my mind. On the day of the big fight – sorry, exam – I'd set three alarms (even though my wired brain would wake me 20 minutes before any of them) and eat a hearty breakfast, before heading to the exam hall where I'd all but shadow-box on the spot, raring to get in there and show what I could do. But most importantly, I would absolutely NOT speak to anyone before the exam itself, for fear of uncovering some titbit of information I'd overlooked and throwing me off my stride. The same applied to immediately after the exam, when I would practically sprint away from the exam hall to avoid the inevitable post-mortem: *What's an ionic bond? Who shot Franz Ferdinand? What's the rule of three?*

Unfortunately, in the age of social media it is utterly impossible to escape the post-exam huddle:

I hope I never get another question on sweets in a maths paper again! #EdexcelMaths

5 years of studying maths in school and all I'll ever remember from it is Hannah's stupid sweets #EdexcelMaths

Hannah eats some sweets. Calculate the circumference of Jupiter using your tracing paper and a rusty spoon. (5 marks) #EdexcelMaths

The eagle-eyed amongst you might recognize those comments as responses to one of the most infamous exam questions in recent history.

#11
There are *n* sweets in a bag.
6 of the sweets are orange.
The rest of the sweets are yellow.

Hannah takes a random sweet from the bag. She eats the sweet.
Hannah then takes another sweet from the bag. She eats the sweet.
The probability that Hannah eats two orange sweets is $\frac{1}{3}$

Show that $n^2 - n - 90 = 0$.

Credit: Pearson/Edexcel

Yes, this is 'Hannah's Sweets', possibly the single best-known GCSE question in British exam history, as set for 16-year-old students by the Edexcel board in 2015. Firstly I'll run through the solution, before the more interesting issue of just how and why this particular sweet-based conundrum became the mother of all viral exam questions and caused an enormous furore, triggering countless news pieces and even a student-led petition to lower grade boundaries.

Before we begin, a quick crash course in some probability basics.* The probability, or likelihood, of an event occurring is represented by some value between 0 and 1, where 0 means impossible and 1 means certain. Often it's useful to represent probabilities as fractions, for example the probability of throwing a 2 when you roll a dice is 1 in 6 or $\frac{1}{6}$, because there is one outcome you 'want' out of six equally likely possibilities.

One more thing: when we want to know the probability of one event *and* another event happening, we multiply the individual probabilities together (as long as the two events in question are independent, in other words one outcome does not affect the other). Say you were throwing a coin and a dice together, and wanted to know the probability of rolling a 2 *and* a head. Their individual probabilities are $\frac{1}{6}$ and $\frac{1}{2}$, so we multiply these together to give a final probability of $\frac{1}{12}$. We can check that this is correct by systematically listing all the possible outcomes (12), of which only one is what we want:

| 1,H | **2,H** | 3,H | 4,H | 5,H | 6,H |
| 1,T | 2,T | 3,T | 4,T | 5,T | 6,T |

---

* If you have recently studied GCSE level Maths or higher, or perhaps you're even a maths teacher, you can probably skip past this next paragraph.

Now let's look at Hannah's sweets. There is a big bag of sweets: 6 orange and the rest yellow. The sum of all the orange and yellow sweets is n, where n is a number greater than 6. The probability of Hannah choosing an orange sweet first is $\frac{6}{n}$. Next she eats the sweet, which at first seems like a non-sequitur, but is actually telling you that the next time she picks there will be one fewer orange sweet, and therefore one fewer sweet in total.

So the probability of the next sweet she picks being orange is $\frac{5}{n-1}$. To find the combined probability (the probability that Hannah eats two orange sweets) we need to multiply these two probabilities, and the question tells us that this combined probability should equal $\frac{1}{3}$:

$$\frac{6}{n} \times \frac{5}{n-1} = \frac{1}{3}$$

The left-hand side can be written as one single fraction:

$$\frac{30}{n(n-1)} = \frac{1}{3}$$

Multiplying both sides by $3n(n-1)$ gives:

$$90 = n(n-1)$$
$$90 = n^2 - n$$
$$0 = n^2 - n - 90$$

After expanding and rearranging we arrive at the required answer of $n^2 - n - 90 = 0$. The algebra in the last few steps may feel unfamiliar to those who haven't done formal mathematics for a few years, but it is not in any way beyond the scope of GCSE mathematics; most 16-year-olds leaving British secondary schools will have carried out these operations regularly. So what's the big deal?

I think this question provides the perfect storm of factors that leads to exam question infamy. Firstly, it merges two areas of mathematics that ordinarily wouldn't have much to do with each other, which is startling to a student (or parent) who is at the edge of their ability dealing with probability *or* algebra, let alone both. It's a bit like seeing your geography teacher in the supermarket: there's no reason for them not to be there, but it feels unsettling to see them in that context and can be momentarily (or permanently) startling.

There's also the quiet-quiet-LOUD aspect of the question, which gives the unexpected algebraic conclusion that extra punch – like a Pixies song. It escalates quickly, as Ron Burgundy from *Anchorman* would say. There's no mention of algebra all the way though and then – bang! – we're hit with a quadratic equation with a big old 90 at the end of it. Certainly, if you have no idea where to start, the intended conclusion is not going to give you any clues.

Finally, and most importantly, this question was set in the age of social media – particularly Twitter – so that reacting to the question with the wittiest meme or gif became almost a competitive sport in itself. Twenty years ago students might have chatted about this question for perhaps a few days, before moving on to more interesting things to talk about. Some students might even have had a friend or relative in another part of the country whom they could phone up for a debrief – *What was that Hannah's sweets question all about?* – and that would be the end of it. But we don't live in that world any more.

As with social media in general, expressing a middle-of-the-road view – *the Hannah's sweets question was interesting but I think I got it* – is never going to get you those sweet likes and retweets. Only the hottest takes survive: that's the rule of social media, whether it's expressing your political opinion or commenting on a question in a maths paper.

Of course it would be ideal if we had a similar question from an exam before the internet was widely available, so that we could compare and contrast the reactions to both. Oh, hang on…

There are n beads in a bag.
6 of the beads are black and all the rest are white.
Heather picks one bead at random from the bag and does not replace it.
She picks a second bead at random from the bag.
The probability that she will pick 2 white beads is $\frac{1}{2}$
Show that $n^2 - 25n + 84 = 0$.

Credit: Pearson/Edexcel

This near-identical question (answer at the back of the book) comes from a 2002 Edexcel GCSE paper, in the very early days of the internet when viral exam questions were a long way off. I'm sure this caused some heated discussion from students as they left the exam hall, but that would have been the end of it. That's why no one has heard of Heather's beads, but mentioning Hannah's sweets in front of a maths teacher is like going into a guitar shop and attempting the intro to 'Stairway to Heaven'.

Try it yourself

Here's an almost identical version of Hannah's sweets for you to try yourself. This adapted version comes from the *Dr Frost Maths* website, the creation of Global Teacher Prize nominee Dr Jamie Frost. The somewhat graphic imagery in the middle part of the question is Dr Frost's own, though I do think it makes it quite clear that the sweet is not replaced!

Neha has n sweets, 3 of which are green. The remainder of the sweets are red.

Neha eats a sweet, does not regurgitate it, and then eats another sweet.

The probability that she eats two green sweets is $\frac{1}{3}$

Show that $n^2 + an + b = 0$, where a and b are constants to be found.

Solve this equation to find how many sweets were in the bag.

Edexcel are reticent to talk publicly about this notorious question, but I was able to convince a covert agent to break ranks and discuss the impact of Hannah's sweets on the exam board and on examinations in general, in the age of social media. I'll call the person *Agent Z* to maintain their anonymity. Was there ever any serious consideration, I wondered, of easing the grade boundaries as a result of this question? Public pressure can of course be quite persuasive.

'Oh no, never,' said Agent Z, glancing in both directions from the park bench where we met. 'It only took a few minutes of checking against the mark schemes to make sure what was asked of students was clear and fair – Hannah's sweets passed that test.

'The furore was a surprise because we hadn't seen anything like this before on social media, but once we realized that a viral exam question could be a thing, this question certainly has the elements needed. Crucially, there was a context students could remember, and a sequence of steps that didn't appear to follow through logically to many students – and they could all share that experience.

'Criticism of GCSE questions at that time, and probably now, is that papers had become too routine, a rehash of past questions testing memory rather than mathematical ability. I now advise teachers to tell their students to expect the unexpected in about 10% of the paper; there will be questions the like of which they have never seen before.' This has had a humorous – if inevitable – consequence though, as Z explained:

'Some teachers have asked if we could supply a list of what these unexpected questions will be so their students can get used to them – missing the point, I fear.'

Does Agent Z have any thoughts on the legacy of Hannah and her sweets?

'I think the question has been good for Pearson/Edexcel – it's almost a brand statement now because Edexcel usually gets a mention whenever it's aired in public. Teachers have told us that their students get it as a rite of passage somewhere along the line, and I hope they continue to.'

The viral 'success' of Hannah's sweets has had another unexpected side effect, in that now every time a maths exam is sat, a cursory glance at Twitter in the hours following the exam will show a flurry of tweets from students desperate to write the funniest response or meme. In return, this seems to have led exam boards to be extra cautious with the questions they set, lest they unwittingly create the next 'Hannah's Sweets' or 'Crocodile and Zebra' furore.

#12 The crocodile and the zebra

A crocodile is stalking prey located 20 metres further upstream on the opposite bank of a river.

Crocodiles travel at different speeds on land and in water.

The time taken for the crocodile to reach its prey can be minimized if it swims to a particular point, *P*, *x* metres upstream on the other side of the river as shown in the diagram.

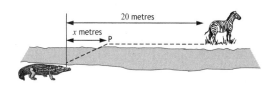

The time taken, *T*, measured in tenths of a second, is given by

$$T(x) = 5\sqrt{36 + x^2} + 4(20 - x)$$

(a) (i) Calculate the time taken if the crocodile does not travel on land. (1)

    (ii) Calculate the time taken if the crocodile swims the shortest distance possible. (1)

(b) Between these two extremes there is one value of *x* which minimizes the time taken. Find this value of *x* and hence calculate the minimum possible time. (8)

This question was set by the SQA board in Scotland during the very same exam period as Hannah and her sweets – summer 2015 – and it caused a similar storm: at one point an article about it was the single most read article on the BBC website.

I'll do the 'easy' parts of the solution here but leave the meaty eight-marker for the back of the book. According to the diagram (and even this part takes some careful thought) if the crocodile does not travel on land at all – swimming directly to the zebra in one straight diagonal line – then the value of x will be 20, because x represents the point on the bank that the crocodile swims to, rather than the horizontal distance that the crocodile crawls. At the other extreme, if the crocodile swims straight across from one side to the other that will minimize swimming, and x will be 0. So parts (i) and (ii) can be solved by swapping x for 20 and 0 respectively, giving answers of 104.40 and 110. However, always be sure to read the question carefully!* T is actually measured in tenths of a second, so the times for the crocodile to reach the zebra are **10.4** seconds and **11** seconds (rounded to three significant figures, as all maths students will tell you is the standard level of accuracy unless you're told otherwise).

This is actually a really nice question about *optimization*. The crocodile wants to optimize the ratio of swimming/crawling to get to the zebra as quickly as possible. Currently it's slightly better to swim directly across the river and then crawl along the whole bank than to swim all the way; although the distance is longer, the crocodile clearly crawls faster than it swims, so it can get to the zebra more quickly. But there may be a 'sweet spot' between the two approaches that gets the crocodile there faster.[†]

Solving this particular question involves a bit of fiddly calculus so I will put it in the back of the book, but I would like to give you a similar

---

* RTQ is the official teacher acronym for this, or RTFQ if you mean 'read the *full* question'.

† If you're thinking that the crocodile could get there quicker by worrying less about algebra and calculus and just getting on with it – stop being such a spoilsport.

optimization problem (and a bit of a classic) that is easier to solve by trial and error. The problem comes in two parts.

#13

(i) A farmer wants to create a rectangular sheep pen, and has 36 m of fencing to use. What dimensions should the pen take to give the sheep as much grazing space as possible?

(ii) She then realizes she has a long barn that she could build her pen against, effectively only requiring three walls to be built. What dimensions should she choose now?

Below is a non-optimal example for each situation, to give you an idea:

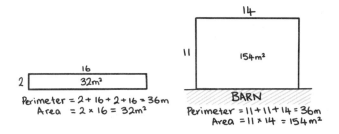

Many people quickly see that the optimal pen for the first situation is a 9 × 9 square, giving a potential area of $81m^2$, but the second case is more interesting. Was your first instinct to also build a square pen against the barn wall? Mine was. But this arrangement requires you to build three 12m walls, creating a 12 × 12 square with an area of $144m^2$. I planted my second example to hint that this wasn't the optimal approach: my 11 × 14 rectangle actually beats this with $154m^2$.

The actual optimal way to build the rectangular sheep pen against a wall is to make it twice as wide as it is long, so that the sides you build are 9m, 9m and 18m, giving an area of $9 \times 18 = 162\text{m}^2$.

Here's a short table showing that this is indeed the optimal arrangement by comparing it to some pens with similar dimensions:

Dimensions (m)	Area (m²)
12×12	144
11×14	154
10×16	160
9×18	162
8×20	160
7×22	154

The symmetry in this table is no coincidence, and I'll also explore this a little in the back of the book. Using a trial and error approach is clearly not the best, but without any better tools we could similarly explore the crocodile and the zebra situation to look for a better solution than those we have so far. The formula for the crocodile's crossing time is:

$$5\sqrt{36 + x^2} + 4(20 - x)$$

Let's plug some values into that formula and see what crossing times we get (remembering to divide by 10 at the end of the calculation):

x (m)	Time (seconds)
0	11
5	9.91
10	9.83
15	10.1
20	10.4

So the 'goldilocks zone', as we might call it, is when x is somewhere around 5–10 metres. Let's zoom in and look a little closer at that region:

x (m)	Time (seconds)
5	9.91
6	9.84
7	9.81
8	9.8
9	9.81
10	9.83

It looks like **8 metres** is the optimal value of x. We could carry on investigating with trial and error, or we could approach the problem algebraically; you'll find this approach at the back of the book. Yes, the zebra has probably escaped by now.

The paper that this question was part of actually *did* have its grade boundaries lowered, but it's unclear from the examiners' report if it was purely because of the crocodile and the zebra. The paper also featured a sixteenth-century problem about a frog and a toad falling down a well – really! – so it might have been that.

#14

A dragon lived in a cave.

The dragon *doubled* in size every day.

Afrer *20 days* the dragon filled the cave.

After how many days did the dragon *half-fill* the cave?

This humorous little gem comes from a Level 6 SATS paper that some 11-year-olds in England sat in 2011. This is probably as close to a 'gotcha' style riddle that you'll ever see in an exam paper,* with the borderline 'trick' answer of **19 days**: if the dragon doubles in size every day, and fills the cave on day 20, then it would have half-filled the cave one day earlier. The inevitably common wrong answer is to say day 10, but this assumes linear dragon growth rather than exponential dragon growth. If you aren't sure what exponential growth means – congratulations! You've either been living in a cave (with a dragon?) during the Covid-19 pandemic, or enough time has passed that you've forgotten all about it. If so, what's the future like? Are there hoverboards yet?

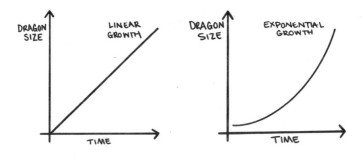

The dragon, contrary to the experience of any living creature in the history of the earth, is not only growing but growing at a rapidly increasing rate. One can't help but wonder what happens to the dragon from day 20 onwards, surely a more interesting situation? This question is actually a direct adaptation of a very well-known riddle that probably originates from China, but with the dragon replaced by a water lily and

---

* With the possible exception of the 'knot' I spent 20 minutes of a university exam trying to assess the complexity of, only to realize it was an unknot. This was in 2005 and it still bugs me to this day.
(Yes, knot theory is a thing. Hence the joke: What's your favourite area of maths? Knot theory – it works better out loud.)

the cave by a pond. Why the question was adapted from lily pond to dragon cave one can only wonder; perhaps dragons and caves are more relatable to 11-year-olds than lilies and ponds?

Exponential growth is actually quite hard to find in everyday life, exactly because increasing rates of growth can't be maintained forever. Cases of a rapidly spreading disease may grow exponentially for a time, but even without methods of prevention the rate will eventually level off when nearly every member of a population has been infected (or died). The earth's population appears to have been growing exponentially for the last couple of hundred years or so as healthcare has improved hugely, meaning people live longer and vastly fewer children die during childbirth or in infancy. But the broad scientific consensus is that as the number of children per family drops worldwide due to wider-spread contraception, the earth's population will settle at around 11 billion over the next century.*

I've long been fascinated by exponential growth. As a teenager, I remember seeing a valentine's card in a shop that said: 'I love you twice as much as yesterday, but half as much as tomorrow.' This is a terrible thing to promise someone! If you lived a long life together and your sweetheart died peacefully on her 100th birthday, the day before this

---

* If the world's population could be represented by water in a bathtub – the water entering through the tap representing people being born, the water exiting through the plughole representing people dying – then for most of human history the water level has remained fairly steady. Within the last few hundred years though, we've found a way to increase the tap's flow (preserving the lives of a greater number of newborn children) while simultaneously narrowing the plughole (improving the health of all people, lengthening lives generally). Disaster! The bath might overflow! Except it won't, because people in most parts of the world are having vastly fewer children than they used to, thanks to contraception and education (bringing the tap's flow right back down). One of my heroes, the recently departed and dearly missed Hans Rosling, was a mastermind at communicating the mathematics behind population growth and levelling. Look for his TED talk entitled 'box by box' for a much better analogy than I have managed here.

would have been the day you loved her with just 50% of your romantic potential. One day before that it would have been 25%, and so on. One year prior, on her 99th birthday, you'd have loved her with 0.0000(… add a hundred more zeros…)0000133% as much as you potentially could. The buyer of this card is essentially promising to hold their beloved in complete contempt for the vast majority of their life together.

I currently have a social experiment running on Twitter, which started in a moment of boredom towards the end of 2019:

Kyle D Evans
@kyledevans

If this gets 1 like I'll continue the thread

(thread)

2:44 PM · Nov 22, 2019 · Twitter Web App

That quickly got a like, so I made it into a thread where I asked for 2 likes, then 4, then 8, etc. When I got to 16, people seemed to need a little extra convincing:

Kyle D Evans
@kyledevans

Replying to @kyledevans

If this gets 16 likes I'll continue the thread and make you a nice cup of tea ☕

4:09 PM · Nov 22, 2019 · Twitter for iPhone

Here's a table showing how long it took me to harvest the required likes for each new step:

Likes	Days after first tweet
1	0
2	0
4	0
8	0
16	0
32	1
64	2
128	96
256	600 and counting

I should confess that I only got the 128 likes by practically begging in front of a hall of over 500 teenagers. This paragraph is essentially me doing the same thing for the 256 likes, but with readers of this book. But the point is the same: exponential growth is very difficult to maintain. At the time of writing there are approximately 320 million Twitter users. Even if every single one of them starts to like my posts, by the 29th iteration there won't be enough Twitter users in existence to keep my thread going. Even if you include all the bots.

Exponential growth truly is a mind-bending concept, no matter how long you've been doing maths or how familiar you think you are with it. When I was in school, the most exciting way to kill time in a boring lesson was to fold a piece of A4 paper as many times as you could.* No matter how hard you tried, it seemed impossible to get past seven folds, even with the most delicate creases and all of your adolescent might put into forcing the last few folds. Why this magic number of seven folds?

---

* I took great enjoyment from one of my old classes, who told me they had never really done this paper folding thing, but they did all play the '2048' game on their smartphones.

It only occurred to me, several years later, to actually look at the maths behind paper folding. A piece of A4 paper is 297 × 210mm, it says so on the front of a ream of paper. A standard ream of printer paper contains 500 sheets, and it's 5cm thick, meaning that 100 sheets are 1cm thick, so that one sheet has a thickness of a hundredth of a cm, or 0.1mm.

Now, every time you fold a piece of paper you either halve the length or the width, but crucially, *you double the thickness with every fold.* So the dimensions of our folded paper at every stage of the first seven folds looks like this:

Fold	Length (mm)	Width (mm)	Thickness (mm)
0	297	210	0.1
1	148.5	210	0.2
2	148.5	105	0.4
3	74.25	105	0.8
4	74.25	52.5	1.6
5	37.125	52.5	3.2
6	37.125	26.25	6.4
7	18.5625	26.25	12.8

When we reach the eighth fold, the thickness would need to be 25.6mm, which is greater than either the length or width would be. Clearly this would be impossible! If your piece of paper could be as long or wide as you need it to be, 45 paper folds would make the thickness of your paper the same distance as from Earth to the Moon.

We must be careful when talking casually about 'exponential' increase though, as it really is incredibly rare in nature. At the time I was writing this, an international news story developed concerning an enormous cargo ship that had got stuck sideways in the Suez Canal, causing a huge pile-up of ships (and memes). One excitable news source declared that the number of queueing ships was increasing *exponentially*. Really? I'm sure the number of ships is increasing, but is the *rate* at which the number of ships is increasing also increasing? If the number of ships truly was increasing exponentially, the manufacturing world would soon have to put all of its efforts into building new ships, simply to keep the Suez queue topped up.

Back to the dragon. Living creatures actually grow in the exact opposite way to our fire-breathing friend, that is, *logarithmically* rather than exponentially. Logarithmic growth means that the rate of growth is positive but always decreasing rather than increasing. For example, humans grow rapidly in their early years and continue to grow up to their late teens or early twenties, but (give or take the odd growth spurt) at a rate that tails off over time.

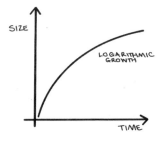

I was once asked during a lull in a pub quiz to explain what exactly logarithmic growth was. After some flubbing and failed hand gestures, I realized that a pub quiz itself was the perfect way to explain. On that night we were the only two members of our pub quiz team present, but we still managed to score nearly as well as when the full regular team of six turn up. Adding an extra team member would have made our team do a bit better – this person might have helped in the food & drink round, where we struggled – but they wouldn't have significantly improved the team's score as there would be an inevitable overlap in our knowledge: most of the things they knew, one of us would have known too. If we rolled in with a team of ten one week, we might only do very marginally better than our usual team of six, since six people with various specialities and interests will cover a huge amount of potential trivia, from kings and queens to classical music. With this in mind:

#15

An orchestra of 120 players takes 40 minutes to play Beethoven's 9th Symphony. How long would it take 60 players to play the symphony?

This rather daft-looking question went viral in October 2017, largely due to the following tweet (which uses a popular meme-y phrase of the time):

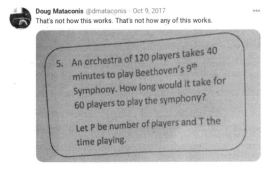

Doug Mataconis @dmataconis · Oct 9, 2017
That's not how this works. That's not how any of this works.

5. An orchestra of 120 players takes 40 minutes to play Beethoven's 9th Symphony. How long would it take for 60 players to play the symphony?

Let P be number of players and T the time playing.

Credit: Twitter/dmataconis

Of course, that's really not how it works at all: the symphony lasts **40 minutes** regardless of how many musicians play it (actually I think the symphony lasts more like 70 minutes, but I did have to look that up. I can never get the classical music questions on *University Challenge*). But, brilliantly, the teacher who actually wrote it fully ten years previously, Claire Longmoor, saw the tweet.

Claire Longmoor #FBPE #remain
@LongmoorClaire

I wrote this!! How did you get this??? I am a maths teacher in Nottingham UK. Wrote this 10 years ago. Here is the original whole worksheet

Credit: Twitter.com/LongmoorClaire

And here is a little more of the whole worksheet:

3. Strawberry Pickers R Us employs 15 people to pick one field of strawberries in 10 hours. How many strawberry pickers do they need to pick one field of strawberries in 3 hours?

   Let T be the time to pick the strawberries and P the number of pickers.

4. Trevor surfs the internet for 2 hours and it costs him £1.20. How much would it cost Trevor to surf for 5 hours and 10 minutes?

   Let T be the time on the internet and C the cost.

5. An orchestra of 120 players takes 40 minutes to play Beethoven's 9th Symphony. How long would it take for 60 players to play the symphony?

   Let P be number of players and T the time playing.

6. Daisy, Bluebell and Maisy Moo the cows eat 200kg of cattle feed per week. Buttercup moves into the farm, how long will the cattle feed last now?

   Let C be the number of cows and T the time the food will last.

It turns out Claire was very much aware of the silliness of the question – in fact, that was the whole point. This question was inserted amongst some serious questions about direct and inverse proportion to test if students were paying attention as they answered the questions, rather than just mindlessly repeating the same process.* Wouldn't you have loved to have had a maths teacher like that? It turns out many people would.

'The overwhelming response was lovely,' says Claire, 'with people saying I had "won the internet" and they wished I was their maths teacher.' Incredibly, this question exploded into life on Twitter just days after Claire created her account, so that one of the first things she ever saw on the site was one of her favourite comedians sharing her isolated maths question from ten years earlier. Her response tweet above garnered 30,000 likes and rising and continues to gather likes and retweets on a daily basis, years later. How does it feel, I wonder, to have such a moment in the sun? Certainly these are numbers that make my cheese pun wilt like a particularly ripe brie.

'I was flabbergasted by the massive response. There were quite a few articles written about me around the world, from *Time* magazine to my local newspaper. I tell all my students and friends and basically anyone who'll listen about my 15 minutes of fame. I still get likes and retweets now over 2 years later.' There is one thing Claire would like to clear up though: 'By the way, I now know that Beethoven's 9th symphony is longer than 40 minutes!'

The shoehorning of direct proportion into places it doesn't belong is fairly common, and usually not intentional, as can be seen in the following example from a 2020 episode of the TV barrel-scraping exercise *I'm a Celebrity, Get Me Out of Here!*

---

* The other answers, if you were wondering, are: **50** strawberry pickers, **£3.10** to surf the internet (who remembers internet cafes?) and **5 and a quarter** days' worth of cattle feed.

#16

> Aled has bought 7 horses.
> Each horse can make his cart go
> 4 miles an hour.
> He has a 336 mile journey ahead.
> How many hours will it take him
> to get to his destination?

I don't actually know what the answer to this question was because I don't watch this show for fear of being hit by a bus the next day and knowing, as I bled to death on the pavement, that this was how I chose to spend some of my last hours on earth. But I'm willing to bet the answer was **12 hours** (336 miles at 28 miles per hour).

Of course this is not how horse-drawn vehicles work. With this model, harnessing enough horses in series can get your cart moving at any speed you require: 18 horses to break the speed limit on UK motorways; about 200 to break the sound barrier; around 168 million horses to travel at the speed of light. Which I suppose would mean you could go backwards in time and write a better question.

In reality, the cart speed would increase logarithmically as you add more horses; increasing with every horse you add, but increasing at a lesser and lesser rate.

You might ask: how come I have such a big problem with this question, but not with the crocodile and the zebra? Aren't they both unrealistic situations that maths has been over-enthusiastically shoehorned into?

My response to this would be that the crocodile and zebra is an honest attempt at a non-trivial optimization situation to stretch and challenge young mathematicians. Animals in the wild actually do follow mathematical models, for example falcons that swoop upon their prey in a logarithmic spiral – the curve shape that means they can keep their eye trained on their prey throughout the descent. The falcon knows no more about logarithms than the average pub quiz team member – probably less – but in trusting its instinct it naturally follows a mathematical pattern. The same could be said of a crocodile attempting to reach its potential dinner as quickly as possible.

The TV question, on the other hand, essentially requires some celebrities to multiply 7 by 4, and then divide 336 by the result (although they could just divide 336 by 7 and then 4, or 4 then 7, since multiplication takes no precedence over division, but we'll come to this shortly). Ideally, we'd also like these calculations to take place in some 'real life' context too. Here's a couple of suggestions:

The classrooms at Byker Primary school have seats set out in 7 rows of 4. There are 336 pupils at the school. How many classes are there?

Anf and Dick earn 7 million pounds for every TV show they present, and they film 4 TV shows per year. How many years would it take them to earn 336 million pounds?

The style of the question can't help but remind me of the following mischievous question trope:

It takes 5 men 5 minutes to dig 5 holes. How long does it take 10 men to dig 10 holes?

This is a classic trick question that wills you in its set-up to go for 10 minutes. Everything about it is pushing you towards that answer: it's like telling someone not to think of a sheep. It becomes instantly impossible to have any other thought. Pause for a second, and the correct answer of **5 minutes** becomes clear: ten men digging ten holes is exactly the same as five men digging five holes – one hole each. So if the first situation takes 5 minutes, so will the second, you'll just have twice as many holes.*

This is a nice example of what's often called a *cognitive reflection* problem, examples of which we'll see in a few other places throughout the book. Every part of you wants to scream out the answer you're being misdirected towards, no matter how hard you try not to. We could certainly all learn a little from this type of problem when it comes to lurching towards quick opinions and hot takes online. Take a step back and a deep breath, and it doesn't take too much perspective to see that our instinctive first response leaves something to be desired.

Apparently though, this classic of the genre is just not quite tricksy enough, as googling words along the line of 'it takes 5 men 5 minutes to dig hole' throws out the following type of riddle:

If it takes 3 men 3 hours to dig 3 holes, how long does it take 5 men to dig half a hole?

You may perform some reasonable calculation, such as realizing it essentially takes a man 3 hours to dig a hole, so 5 men could dig a hole in three-fifths of an hour, or 36 minutes, therefore 5 men could dig half a hole in 18 minutes. But the answer to these questions seems

---

* If you're wondering what exactly constitutes a hole, the officially recognized size of a hole, defined in 1967, is one ten-thousandth of the volume of the Royal Albert Hall.

to be: there's no such thing as half a hole. Yawn. Some people are just determined to ruin all the fun.

Here's another regularly occurring viral problem that originates from the classroom. I'm rather fond of it:

#17

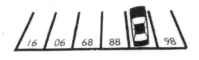

(Source unknown)

What space number is the car parked in?

I love this little nugget; it makes use of a certain property of numerals that is puzzling at first, but, once you see it, suddenly becomes very obvious. It's the kind of problem that you can imagine was written when someone actually walked past some painted car park numbers and spotted the feature that has been exploited. I've actually argued – good-naturedly – with mathematicians and puzzle fans who claim this is *not* a good puzzle because it involves more observation and lateral thinking than mathematical deduction. But I'm sticking to my guns with this one. I think it's great and it's one of the first puzzles I stuck to my front door when I was running a weekly puzzle for local kids during Covid-19 lockdowns. (The answer is **87**; the numbers appear upside down, readable to the person driving into the space.)

An interesting side note: here's the full image that I found when I looked up 'car park logic puzzle':

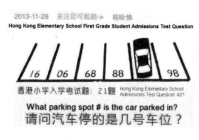

(Source unknown)

The Cantonese text asks for a response in under 20 seconds. Now, although I'm a big fan of this question, I'm not sure if I'd have been able to answer it in 20 seconds the first time I saw it. Apparently this question was set to first-graders in Hong Kong, aged 6–7. Any problem seems to gain an extra-viral push in the West if it was originally set to Asian schoolchildren, for reasons that we could spend half a book unpacking and exploring. One thing I can say for certain is that it led to some very obvious headlines:

Are YOU smarter than a Hong Kong first grader? Logic puzzle from Chinese primary school exam stumps adults but 6-year-olds can crack it in just 20 seconds

Perhaps the most infamous of this category of viral puzzle is 'Cheryl's birthday', which spread like wildfire in 2015, and is one of the few questions in this book to have its own Wikipedia page:

#18

Albert and Bernard just became friends with Cheryl, and they want to know when her birthday is. Cheryl gives them a list of 10 possible dates.

15 May	16 May	19 May
17 June	18 June	
14 July	16 July	
14 August	15 August	17 August

Cheryl then tells Albert and Bernard separately the month and day of her birthday respectively.

Albert: I don't know when Cheryl's birthday is, but I know that Bernard does not know either.
Bernard: At first I didn't know when Cheryl's birthday is, but I know now.
Albert: Then I also know when Cheryl's birthday is.

So when is Cheryl's birthday?

This question was posted on Facebook by Singapore TV presenter Kenneth Kong, suggesting that it was for primary year 5 children, ten-year-olds. What the hell? I honestly had to read this problem five or six times to even understand it! As it happens, it later transpired that it actually formed part of a Singapore and Asian Schools Math Olympiad test paper, aimed at particularly able 14- and 15-year-olds. That feels a bit more reasonable, though it's still pretty damned tricky.

If you're able to hold all of this information in your head and come to the correct answer without putting pen to paper, I bow to your Sherlock-esque mind palace. For me, it was essential to draw a grid of all the potential birthdays and then remove wrong answers as they became apparent:

May		15	16			19
June				17	18	
July	14		16			
August	14	15		17		

Albert: I don't know when Cheryl's birthday is, but I know that Bernard does not know either.

Let's slow this down. Cheryl whispers a month to Albert and a number to Bernard. Albert says 'I don't know when Cheryl's birthday is', but of course he doesn't; there's no month Cheryl could have whispered that would have locked down the full date, because every month has more than one possible date.

However, whatever Cheryl whispers to Albert is enough for him to realize that Bernard doesn't yet know the answer either. If Cheryl told Albert 'May' or 'June', then she *could* have told Bernard '18' or '19', and in doing so given Bernard the full answer (if she said '18' that could only mean 18 June, '19' could only mean 19 May). She did *not* give Bernard the full answer, so Albert knows she *definitely* did not whisper '18' or '19' to Bernard, so she must have whispered 'July' or 'August' to Albert.

May		15̶	16̶			19̶
June				17̶	18̶	
July	14		16			
August	14	15		17		

Bernard: At first I didn't know when Cheryl's birthday is, but I know now.

Bernard didn't know the full date from what Cheryl whispered to him (so, as we've established, she didn't whisper '18' or '19') but, once Albert speaks, Bernard does know the full date. From this we can say for certain that Bernard was not told '14' – if he had been, then he still would be no closer to solving the problem (the solution could be 14 July or 14 August.)

May		15̶	16̶			19̶
June				17̶	18̶	
July	1̶4̶		16			
August	1̶4̶	15		17		

Albert: Then I also know when Cheryl's birthday is.

The only dates remaining are 16 July, 15 August or 17 August, and Albert now knows, as a result of Bernard's last response, the full date of Cheryl's birthday. If Albert had been told August, removing the possibility of the 14th would still leave him in the dark (as 15 August or 17 August could still be the answer). Therefore, the only date that can work is **16 July**.

All good maths teachers will encourage their students to check their answers, so let's take the date of 16 July and run through the conversation between Albert and Bernard once more, just to be sure:

Albert: I don't know when Cheryl's birthday is, but I know that Bernard doesn't know either.

Albert is told July, so he knows Bernard was told 14 or 16. He therefore knows that Bernard couldn't yet know, because 14 and 16 are non-unique in the list of dates.

Bernard: At first I didn't know when Cheryl's birthday is, but I know now.

Bernard has a 16, which wasn't enough at first to plump for 16 July because 16 May was an option. However, when Bernard sees what Albert is thinking, he realizes Albert has ruled out May and June, so Bernard can settle on 16 July.

Albert: Then I also know when Cheryl's birthday is.

Once Albert knows that Bernard knows the full answer, he knows that Bernard can't have been told '14', and his only options are 14 July and 16 July, so he too can plump for 16 July.

Phew! What a workout. This problem was pushed into extra-viral territory when quite a substantial number of online puzzlers found apparently legitimate reason to settle upon the answer of 17 August, leading to much bickering and indignation with few people willing to see the other side's point of view.

Eventually the original Singaporean setters of the question confirmed that the intended solution was indeed 16 July, and explained where the error in the contrary approach lies. It's possible to find the 17 August justification online, but don't pay it too much attention because 17 August truthers are idiot sub-humans who frankly don't deserve to share oxygen with the rest of us (I'm joking!).

If you haven't come across the 'I don't know the answer... but now I do' type of puzzle before, may I recommend the following problem about a bizarre prison:

#19

Four prisoners are standing in the following way, so that each of
  them is wearing either a black or white hat and facing a wall.

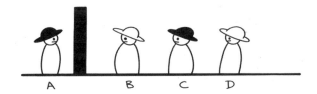

The prisoners know that there are four of them, and two of each
  colour of hat. They cannot turn around or remove their hat. If any
  prisoner can declare the colour of their hat out loud, then all the
  prisoners will be released. Can any prisoner correctly declare their
  hat colour?

It looks at first as if all the prisoners are doomed to eternity in this bizarre institution. Prisoner A is clearly stuck, since they are staring at the wall with no other prisoner to base their decision on. Likewise prisoner B has no way to ascertain their hat colour and would be purely guessing. Prisoner C sees a single white hat ahead of them, so that's no use, and prisoner D sees a black and a white hat: also no use. If prisoner D saw, say, two black hats, they would be able to declare their own hat to be white, but alas no luck.

Hang on though… if prisoner C is thinking along the same lines that we are, then prisoner C knows that prisoner D is not looking at two hats the same colour. Once they have waited for a few seconds in silence, prisoner C can deduce from the silence in the room that the prisoner behind them is looking at one black and one white hat, and therefore prisoner C knows that their own hat is the opposite colour to prisoner B's hat. **Prisoner C can declare 'black!'**

This prisoner dilemma was the first thing I thought of when I saw Cheryl's birthday problem; it's rare that a problem asks you to put yourself in someone's shoes and then consider how your understanding would change over the passage of time.

Riddles that involve prisoners having to crack mathematical puzzles to free themselves, or often the entire prison population,* are fairly common and can vary quite enormously in difficulty. Here are three you might like to try, increasing in difficulty (in my opinion). Answers at the back of the book:

1. A prison has 100 inmates and 100 guards. Each prisoner is in their own cell, and the prisoners, cells and guards are all labelled 1–100. While the prisoners are asleep, guard 1 turns the lock on every cell (so that every cell is now unlocked). Guard 2 then turns the lock on every second lock, so that locks 2, 4, 6, etc. are now locked again.

---

* This is why puzzle enthusiasts should never be put in charge of high-security prisons.

Guard 3 then turns every third lock, locking some cells but unlocking others. This continues until all guards have had their turn. In the morning, some prisoners will find their cells unlocked, and freedom theirs! But which?

2. A prison has ten prisoners, all in solitary cells, and an eleventh empty cell. The empty cell has nothing in it but a switch that does nothing, and starts in the 'down' position. The warden explains that every night one prisoner will be chosen randomly, woken while the others sleep, and taken to the empty room. They'll then return to their cell. The warden will release all prisoners as soon as any prisoner can correctly announce that all ten prisoners have visited the room. If they announce incorrectly, their opportunity is lost. The prisoners are given five minutes to decide on their strategy, after this they will never be able to speak to each other again. What should their strategy be?

3. There are ten prisoners in a queue, so that each prisoner can see everyone in front of them but no one behind them. Each prisoner is wearing a hat, either red or blue, though there are not necessarily five of each. The prison guard starts with the prisoner at the back of the queue and asks them to declare what colour of hat they are wearing. If they get it right – you guessed it – they'll be released. If they get it wrong, they remain in jail for life. The prisoners are allowed to talk tactics first, but not once the calling has begun. What should they do to maximize their chances of success?

The tendency of unusual exam questions to go viral has led to a new trend of children's difficult homework questions becoming minor news stories. At times, incredibly standard homework questions have become national news (on *very* slow news days). The following comes from the website of English newspaper *The Mirror* (the sixth most popular in the country) in late 2020:

Mum stumped by daughter's maths homework begs fellow parents for help.

The mum took to Facebook after her daughter's tough maths homework baffled her, but the tricky question has everyone scratching their heads.

Source: https://www.mirror.co.uk/lifestyle/family/mum-stumped-daughters-math-homework
-22814891

What was the question in question? Read on.

#20
Write in index form:

(a) $\dfrac{1}{x^2}$

(b) $(\sqrt[3]{x})^2$

I can't emphasize this strongly enough for anyone who isn't au fait with mathematics: these questions are absolutely standard fare for a pupil aged around 14–16. Completely normal. The article goes on to quote a few people's Facebook responses:

A popular solution among members was that the answer is 'x to the power of 2 over 3', but this confused people more.

'If I wasn't confused enough reading the question I'm now even more confused reading the answers?! Power?!?! 2/3 of an x !!! But what the hell is x?!' exclaimed one person.

This is news, apparently. Someone suggesting the correct answer, and someone else exclaiming that they find the correct answer confusing.

May I suggest a few alternative headlines?

Woman Can't Remember Something She May Have Been Told 25 Years Ago

People Find it Hard to Explain Mathematics Online

Maths is Hard

Don't get me wrong, I'm very happy to see popular mathematics stories in the everyday news, but this simply isn't a story! This kind of situation – a parent finding a child's homework difficult – happens in thousands of homes around the country every evening. All this kind of story does is glorify anti-intellectualism and reinforce a sort of unspoken perceived pointlessness of algebra, or mathematics in general. You'll see something similar when adults declare their consternation that primary age children have to learn about split digraphs, subordinate clauses and fronted adverbials. Guess what? The kids can manage all this just fine! Sure, perhaps we don't need to put the jargon front and centre, but the actual content is perfectly manageable. Often it's adults telling kids (or implying) that something's too hard that makes them think it is.

It only takes a surface-level understanding of how social media works to see why this kind of story has become a popular trope. As you'll surely have experienced, articles like this are meant to pique our interest and make us want to have a go, then to post our solution, increasing the viral effect. This particular example is a complete failure of a story, in that the maths involved is very specific and excludes most readers instantly.* But, generally, maths stories or puzzles that gain a lot of traction online are those that entice the reader to have a go at some maths, and ideally to argue with each other about it too (since petty argument seems to be the currency of the modern world).

Here's an example of a viral homework question that I find much more interesting, from just before this book went to print:

#21
True or False?
This shape has two right angles.

Explain your answer.

Credit: www.whiterosemaths.com

This was posted by maths lecturer and pandemic homeschooler Dr Kit Yates, with the caption: 'This was my daughter's (7) maths homework on Monday. Can someone help me out with the answer?'

One line of thought says that an angle can only be formed where two straight lines meet, so that, no matter how much something *looks* like a right angle, if it's formed by a curved line and a straight line, it cannot be. This is how schoolchildren would first learn about an angle, so the 'correct' answer the homework was looking for is '**false**'.

---

* The answers are x^{-2} and $x^{2/3}$, which is obvious if you've recently studied GCSE maths, and meaningless if you haven't. The method behind the first answer is at the back of the book, if you're interested.

On the other hand, as Dr Yates pointed out when interviewed on TV,* imagine walking around the centre circle on a football pitch. At any point, if you stop and turn 90 degrees inwards you will find yourself pointing directly towards the centre spot. In mathematical terms, we would say that a radius of a circle always meets a tangent at 90 degrees.

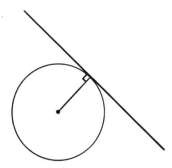

So if someone put a gun to your head and forced you to say there was an angle at each side of the semicircle (an unlikely situation, granted), you would have to say that it's a right angle. If you were to zoom in to the micro-level – and plenty of people on Twitter certainly tried to – you would see that to all intents and purposes the angle at either end is a right angle.

Which is it then: two right angles or no right angles? Two or none? You have to decide! This question is in so many ways the quintessential piece of viral maths, having all of the following key features:

1. An apparently binary right/wrong set-up. It's right there in black and white: two right angles – true or false? Immediately we're encouraged to pick a side and stick to it.

---

* BBC *Points West*, no less. What do you mean, you didn't see it?

2. The plea for help. By asking 'can someone help me out?', Yates is literally inviting the internet to have their twopence worth. This includes many hundreds of people who take the request literally at face value with a short 'false'. Let's be clear: Yates is not actually asking for help because he doesn't know the answer; he's a maths lecturer who will have a grasp of fairly basic geometry. Of course what he's really saying is: *let's discuss how this is actually quite a nuanced question to ask a seven-year-old.* But asking 'can anyone help…' – whether genuinely or with a knowing wink – is a direct route to opening up the reply floodgates.

3. The 'it's all changed since my day' angle. As we've seen previously, all parents can empathize with struggling to help with their children's maths homework.* There's something so seductive about children being able to understand something – especially maths – better than their parents, as seen in recent TV shows such as *Are You Smarter Than a 10 Year Old?*

Throw all this together and you have a perfect viral maths melting pot. Yates himself thinks it's the last of the above points that got the media interested: 'I think they picked it up because they saw a maths lecturer being "baffled" by their daughter's problem, but of course they missed some of the nuance in the question and some of the comments I'd posted below the original tweet explaining the subtleties in the problem.

'As well as people who simplified the problem and wouldn't entertain any angle not between two straight lines, there were other people who went the other way and started talking about non-Euclidean geometry,

---

* The Disney Pixar film *Incredibles 2* contains a wonderful sequence in which Mr Incredible perseveres for hours on end to understand his superhero son's homework, a true role model for us all.

probably in an attempt to show off. In short, I think this question is a great jumping-off point for people to talk about different areas of maths, but probably not the best or most definitive example to test a seven-year-old's understanding of angles.'

Which I'm sure you'll agree is a perfectly logical, reasonable and utterly non-sensational and non-newsworthy conclusion to come to. What the media wants is: yes or no; left or right; right or wrong? In a humorous, if inevitable, conclusion to the aforementioned interview, after explaining the nuance of the problem carefully and quite clearly, Yates was asked by the TV presenter: 'So what is it then: true or false?' Yates' response? 'Yes.'

It's rather infuriating that this kind of false binary choice question makes up a large proportion of maths in the mainstream media. I wonder if we could all do a bit better? Perhaps an inspirational story about a young person doing some brilliant maths, that also involves maths simple enough for the layperson to have a go at and appreciate the beauty in it? First a little bit of pre-work...

#22
Which of these numbers is *not* divisible by 5?   25   50   18   155

We'll start simple. It has probably occurred to you that every number in the 5 times table ends in a 5 or a 0. So, to check whether a number is divisible by 5, you simply need to look at the last digit and check if it's a 5 or a 0. So the odd one out is **18**.

Which of these numbers is *not* divisible by 3?
63   147   717   14,701

The 3 times table does not recur in such a satisfying way as the 5 times table, so we need to think a bit harder here: 63 is 3 more than 60, and 60 is divisible by 3, so 63 must be too. 147 is 3 less than 150, which is divisible by 3, so 147 similarly passes the test. Notice that 14,701 is very similar to 147 – in fact it is just over 100 times bigger. 14,700 is divisible by 3, since it's 100 lots of something that is divisible by 3, so **14,701** fails the test.

You may, however, have remembered the quicker way from back when we were doing 9 times table tricks, that is, to take the digital root of the number, and if that digital root is divisible by 3, then the number itself is divisible by 3:

63	147	717	14701
6 + 3 = 9	1 + 4 + 7 = 12	7 + 1 + 7 = 15	1 + 4 + 7 + 0 + 1 = 13
✓	1 + 2 = 3	1 + 5 = 6	1 + 3 = 4
	✓	✓	✗

Which of these numbers is *not* divisible by 9?
99   891   798   9018

Again, there are some quick clues you might be able to pick up on: 99 is familiar as eleven 9s; 891 is 9 less than 900, so that will pass the test. But there is of course the divisibility rule from the first chapter, which states that if the digital root of a number is 9, the number itself is divisible by 9.

99	891	798	9918
$9 + 9 = 18$	$8 + 9 + 1 = 18$	$7 + 9 + 8 = 24$	$9 + 9 + 1 + 8 = 27$
$1 + 8 = 9$	$1 + 8 = 9$	$2 + 4 = 6$	$2 + 7 = 9$
✓	✓	✗	✓

So **798** is not divisible by 9.

Which of these numbers is *not* divisible by 11?
99   242   951   33,011

Things get a bit trickier here if we don't know a rule. But as always there are some clues to latch on to: we have our old friend 99 again, and you might get good vibes about 242. Note that 242 is 220 + 22, which is ten lots of 22 plus another 22, in other words eleven lots of 22.*

Can either of the other candidates be broken down into ten lots of a value plus the value itself? 33,011 has a pleasing symmetry, and after a bit of playing around we can see that it's 30,010 + 3001, in other words 11 lots of 3001. The odd one out is **951**.

There is a rule for divisibility by 11, but it's a little harder to remember than the previous rules: take the alternating sum of the digits in the number (subtract then add then subtract then add…), reading from left to right. If that is divisible by 11 (including zero), then so is the original number. For example, 605 is divisible by 11, because $6 - 0 + 5 = 11$, and so is 9834, because $9 - 8 + 3 - 4 = 0$.

99	242	951	33,011
$9 - 9 = 0$	$2 - 4 + 2 = 0$	$9 - 5 + 1 = 5$	$3 - 3 + 0 - 1 + 1 = 0$
✓	✓	✗	✓

---

* You might instead have noticed that 242 is two lots of 121, and 121 is 11^2, so it passes for that reason.

So **951** is not divisible by 11.

Which of these numbers is *not* divisible by 7?
 184   532   987   133

The divisibility rule for 7 is rather complicated. In fact, it's so fiddly that in some maths books it's left out altogether – this was the case when 12-year-old British schoolchild Chika Ofili was handed a problem-solving textbook to keep him busy during the summer holiday in 2019. I'll let his teacher, Mary Ellis, pick up the story.

'There was no test listed for checking divisibility by 7. The reason why it was missing is because there is no easy or memorable test for dividing by 7, or so I thought! In a bored moment, Chika had turned his mind to the problem and this is what he came up with. He realized that if you take the last digit of any whole number, multiply it by 5 and then add this to the remaining part of the number, you will get a new number. And it turns out that if this new number is divisible by 7, then the original number is divisible by 7. What an easy test!'

For example, take the number 532.
Remove the last digit and multiply it by 5: $2 \times 5 = 10$
Then add this to the remaining part of the number: $53 + 10 = 63$
At this stage you can either spot that 63 is divisible by 7, so 532 must
   be too, or if you needed to you could go a step further:
   $6 + 3 \times 5 = 6 + 15 = 21$. 21 is divisible by 7, so 532 is too.

Essentially, if the number you started with is divisible by 7, then every time you perform 'Chika's Test', the new number will remain divisible by 7. The first time you spot a number along the way that's divisible by 7, you know that every number in the chain was too.

Take the number 987.

Step 1: $98 + 7 \times 5 = 133$. Is this divisible by 7? If you're unsure, go again:

Step 2: $13 + 3 \times 5 = 28$. 28 is definitely 4×7, so every number we met along the way, i.e. 987 and 133, is divisible by 7.

By process of elimination this means that **184** must be the only number not divisible by 7, and Chika's Test shows this too:

$$184$$
$$18 + 4 \times 5 = 38$$
$$3 + 8 \times 5 = 43$$
$$4 + 3 \times 5 = 19$$

Depending on your memory of the 7 times table, you may have been able to stop at 38, realizing this is not a multiple of 7, but if not you can take the test as far as you need to. I decided to stop at 19, which is clearly not a multiple of 7, so neither are 43, 38 or 184.

Chika and his classmates interrogated his new rule and couldn't find a single counter-example, as Chika himself explains: 'My classmates were very surprised; most of them couldn't believe it was even true. It felt good knowing that I had discovered something that could make life easier for people across the world.' In fact, one can prove that Chika's Test will always work and never give a false positive.*

As it happens, you could start by removing the last digit of a number and multiplying it by 5, 12, 19, 26, 33, ... (anything that is *5 modulo 7*,

---

* There's a lovely proof by Simon Ellis – brother of Mary – at www.simonellismaths. com/post/new-maths

which means 'five more than a multiple of 7') and that would work too. Or, you can multiply the last digit by 2, 9, 16, 23, … (anything that is *2 modulo 7*) and then subtract instead of adding.*

The double-the-last-digit-and-subtract method is by far the most common divisibility test for 7 that you'll find on the internet, but it does have some pitfalls. Observe what happens when we apply this test to 532:

$$532$$
$$53 - 2 \times 2 = 49$$

At this stage, if you didn't spot by eye that 49 is divisible by 7, you'd have to go one step further. This is where things get a bit wonky:

$$4 - 2 \times 9 = -14$$

We do have something that's divisible by 7, but we've gone into negative numbers, which doesn't feel ideal. Chika's Test will never stray into negative numbers, and it also sticks to addition, which tends to be easier to carry out mentally than subtraction. For these reasons, many people (including myself) find Chika's Test more satisfying than the double-and-subtract method, and it has started sneaking into a position of greater prominence in online listings of divisibility tests. Some nay-sayers pointed out that Chika's method wasn't 'new' at all; it had been around for years and just wasn't as common as the other method. To these people I say: does Soft Cell's version of 'Tainted Love' become less enjoyable when you find out it's a cover version of a song that had been around for ages? No, it does not.

---

* Those with a little more mathematical training might spot that adding 5mod 7 is the same as subtracting 2mod 7 anyway, if the aim is to maintain divisibility by 7.

Chika's discovery, it seems, has lit a fire in the young man: 'It has inspired me to continue studying maths to higher level. I want to find more confusing and difficult problems that I could solve, and make even more discoveries.'

What a fantastic story of an inquisitive young mind, and the maths involved doesn't contain any long-winded formulae or unwelcoming symbols. Surely this would be a shoo-in for any news outlet worth their salt, certainly ahead of trashy articles about students' difficult homework? Unfortunately, though Chika's story made a reasonable-size splash in the world of maths news, I can't find a mention in any major news outlet. Wouldn't it be wonderful to see more maths stories like this in the news? One can but hope.

4

OUT OF ORDER

The trouble with BODMAS

#23:
Calculate 60 + 60 × 0 + 1.

It's fair to say that, short of Brexit, Trump and whether the jam or the cream goes first on a scone, there's very little in the history of social media that has the ability to stir up trouble like maths problems that involve using operations in the correct order. The above problem – perhaps with different numbers involved – has gone viral several times in recent years, and almost certainly will continue to do so.

Any post of the above problem on social media will inevitably be inundated with replies falling into one of two answer categories: 1 or 61. Which did you answer with? This problem has been created to cause a rift between those who do or don't have a recent mathematical education: those who don't will likely answer 1, calculating from left to right; those who do will correctly answer **61**, carrying out the multiplication first.

It's also very likely that comment threads following this kind of question will be flooded with the acronym BODMAS, BIDMAS or PEMDAS, often declared with an argument-closing air of finality. These acronyms are taught fairly universally in English-speaking schools to help students remember the correct order in which mathematical operations should be carried out, to remove any ambiguity from problems such as that above.

Brackets Order   Division Multiplication Addition Subtraction
Brackets Indices Division Multiplication Addition Subtraction
Parentheses Exponents Multiplication Division Addition Subtraction

I don't think I need to explain the last four letters in these acronyms, but perhaps the first two:

Brackets / Parentheses: These little symbols ( ) are very useful for writing smiley faces in text messages if you were born before 1990, but also for separating text from other text, or mathematical symbols from other mathematical symbols. Just as commas and apostrophes can drastically change the meaning of a sentence,* adding brackets can adjust the intended meaning of a string of mathematical operations.

Order / Index / Exponent: All of these refer to the mathematical notation for repeated multiplication, that is, where 4^3 means $4 \times 4 \times 4 = 64$, multiplying together a string of three 4s.

So, in any string of mathematical operations, we should deal with brackets or parentheses first, then any indices or exponents, then work through division and multiplication, before finally dealing with addition and subtraction.

For example, there are no indices or exponents in the above problem,

---

* Good grammar, after all, is the difference between knowing your shit and knowing you're shit.

but the insertion of brackets could very easily be used to change the intended solution:

$(60 + 60) \times (0 + 1) = 120$
(brackets first, then multiply 120 by 1)

$(60 + 60) \times 0 + 1 = 1$
(brackets first to give 120, multiply that by 0, then add 1)

$60 + (60 \times 0) + 1 = 61$
(brackets first to give 0, then add that to 60 and then to 1)

Only in the last of those three examples would the brackets not be required, since the BODMAS rule implies that the multiplication in the middle of the sum should be carried out before the addition on either side, as follows:

$60 + 60 \times 0 + 1 = 60 + 0 + 1 = 61$

Why have a BODMAS rule at all, you might ask? Well, it isn't always advantageous to write lots of brackets in your mathematical commands, especially if you are writing computer code, where efficiency is everything. So an order of precedence is a natural step to take to impose some order into potentially ambiguous commands. That's not to say that it isn't a controversial subject, as we shall see.

Try it yourself

1) $20 - 4 \times 2$
2) $16 \div 2 + 6$
3) $16 \div (2 + 6)$
4) 2×5^2
5) $(2 \times 5)^2$

#24

Question for the over 50s. When you were at school, what would the correct answer to this maths question be: $7 + 7 \times 3$?

A: 28   B: 42

This exact question was tweeted by my former childhood footballing hero Matthew Le Tissier during the great Covid-19 lockdown of 2020, in which some celebrities became so bored that they began to poll their followers on basic order of operations questions. This particular tweet received 20,000 votes in just a few hours, with 56.8% of people correctly answering **A: 28**, and the other 43.2% of people misremembering their 1970s education, perhaps permissibly.

THE WORST TRICK IN THE BOOK

While we're on the subject of overexcitable celebrity tweeters, here's without doubt the worst trick in the whole book, as shared in summer 2020 by a celebrity whose identity I will protect by renaming them Sir Adam Sucrose:

Today is a very special day! There's only one chance every thousand years...

Your age this year + your year of birth = 2020

It's so strange that even experts can't explain it!

Far be it for me to suggest that I am any kind of expert, but I do believe it's actually the very definition of how age is calculated that if you add your age to your birth year you will get to the current year, be it 2020, 2049, 1066 or any other year. Oh, except it only works for people who have already had their birthday that year, so when Sir Adam had posted this in late May, it wouldn't even have worked for half of his 5 million followers.

Are you getting the hang of order of operations now? Let's try something a little harder:

#25
What number should replace the question mark?

$20 - 4 \times 2 + 8 \div 2 + 6 = ?$

First let's run through the most forgivable wrong answer, which would be to work from left to right, giving:

$20 - 4 = 16$

$16 \times 2 = 32$

$32 + 8 = 40$

$40 \div 2 = 20$

$20 + 6 = 26$

If the instructions were read out loud this is almost certainly the answer you'd come to, since memorizing the whole string of calculations and then prioritizing multiplication and division would be extremely demanding (more of this later). Since the instructions were written and delivered all in one go, we can prioritize multiplication and division to give the correct answer of **22**:

$$20 - 4 \times 2 + 8 \div 2 + 6$$
$$= 20 - 8 + 4 + 6$$
$$= 22$$

If, however, you were playing the family quiz board game from which this question has been taken, the intended answer is... 2. I'll let them explain:

Answer: 2
Order of calculations is $\times \div + -$
i.e. $20 - (4 \times 2) + (8 \div 2) + 6 = 2$

Dear reader, award yourself an extra point if you can see the mistake the family quiz game has made in arriving upon the incorrect answer of 2. After all, it seems that they have made the correct first step, in prioritizing multiplication and division as shown in their workings above.

The key to their error, and this often happens when BODMAS is used over-enthusiastically, is that addition takes no precedence over subtraction: they take equal precedence. So although the first step in the solution is spot-on:

$$20 - 4 \times 2 + 8 \div 2 + 6 = 20 - (4 \times 2) + (8 \div 2) + 6$$

And this is then correctly simplified to the following:

$$20 - 8 + 4 + 6$$

The error comes in the next line. The solution interprets BODMAS as meaning addition must come before subtraction, so next the 8, 4 and 6 are added, before subtracting this total from 20 to incorrectly come to an answer of 2. But, when applying BODMAS, it is important to understand that division and multiplication have *equal* precedence, and addition and subtraction have *equal* precedence. For this reason, some teachers now adopt acronyms such as BOMA or BIMA:

Brackets, **I**ndices, **M**ultiplication or Division, **A**ddition or Subtraction

Earlier I mentioned that multiplication and division are the same operation, because every multiplication operation could be equally performed as division (multiplication by $\frac{1}{2}$ is the same as division by 2). Similarly, there is no real difference between addition and subtraction (subtracting 5 is the same as adding –5). For this reason, multiplication and division take *equal* precedence, as do addition and subtraction. It is meaningless to prioritize addition over subtraction or vice versa, since they are the same thing!

The observant reader may even have noticed that the most common American acronym for order of operation, PEMDAS, does not even

Some prefer the acronym GEMS for order of operations. *GEMS??* What's going on there? Let me explain:

Grouping, **E**xponents, **M**ultiplication, **S**ubtraction

'Grouping' is being used here instead of 'brackets' or 'parentheses' because sometimes there are 'assumed' brackets that less experienced mathematicians would not necessarily regard as such. Take the following example:

$$\frac{3+5}{2} \times 5$$

By mathematical convention there is no 'need' for brackets on the '3 + 5', but this is where we should naturally start because the 3 and 5 have been grouped together by placing them on top of the fraction bar.

Using 'GEMS' also makes use of a neat trick I only discovered recently: if you always subtract before adding you can never go wrong, even if you slightly misunderstand how order of operation rules work. Let's have another look at the board game question from above.

$20 - 4 \times 2 + 8 \div 2 + 6 = ?$	Evaluate the multiplication and division 'groups'...
$20 - 8 + 4 + 6 = ?$	*Subtraction first...*
$12 + 4 + 6 = ?$	Add to finish...
$12 + 4 + 6 = 22$	

Subtraction has no precedence over addition, but if you treat it like it does, this 'belt and braces' approach will make extra sure you don't make a mistake.

have the last four letters in the same order as BODMAS or BIDMAS:[*] Since the M and D are describing essentially the same operation, as are the A and S, it follows that PEMDAS and BIDMAS can still suggest

---

[*] This is probably because the phrase 'Please Excuse My Dear Aunt Sally' is often used as a mnemonic device. I suggest 'Please Excuse My Aunt' works equally well, and puts less guilt on the shoulders of poor Sally for her unexplained sin.

the same order of operations. It's worth noting at this point that non-English-speaking countries do not seem to have equivalent acronyms for order of operations, which seems to lead to better understanding and less rote learning. In German there is a lovely phrase, *Potenz vor Punkt vor Strich*, meaning something like 'powers before multiplication/division before addition/subtraction'. *Punkt*, literally meaning 'point' or 'dot', can stand for either multiplication or division, while *strich*, literally meaning 'line' or 'dash', means addition or subtraction. I love this and will be using it from this *punkt* onwards.

It's worth noting at this point, however, that BODMAS, BIDMAS, PEMDAS or GEMS are merely a convention that the whole mathematical community has decided to agree upon, for many and various good reasons, but still just that: a convention. There will undoubtedly be occasions when it is not as appropriate to use the standard BODMAS rule, such as the following.

#26
Here's a numerical problem
For all of you to try to get right
The start of the equation is seven minus five
Now multiply the answer that you have by nine
Take away a six and then add on an eight
Divide it all by four
And now just calculate
The answer to this question with the power of your brain

This doesn't look like it's going to cause any order of operations issues, and we should all comfortably agree on the answer of **5**. This problem comes from *Chris Moyles' Quiz Night*, a British television programme that ran from 2009 to 2012 and featured a weekly slot where pop stars such as One Direction, McFly and Alphabeat would change the lyrics of a well-known song to a mathematical problem to be solved live (the above question comes from an adaptation of the song 'Everybody's Changing' by the band Keane – if you know it, the chorus kicks in on the words 'Now multiply...')

BODMAS begins to rear its ugly head when you use these clips in the classroom, as I myself found out when attempting to spice up my maths lessons at the time. Students would write the problem across the page so that they could find the answer in their own time without the pressure of doing live calculations while listening to the song:

$$7 - 5 \times 9 - 6 + 8 \div 4$$

I would then have to be careful to reward both the students who had worked from left to right, as is clearly the intention here, but also those who had used BODMAS on what they had written down!

$$7 - 5 \times 9 - 6 + 8 \div 4$$
$$= 7 - 45 - 6 + 2$$
$$= -42$$

Perhaps an important lesson here is that if the students correctly 'punctuated' their notes with brackets, they would also reach the intended answer of 5 with no ambiguity:

$$(((((7 - 5) \times 9) - 6) + 8) \div 4) = 5$$

I'm sure you agree that all those nested brackets look quite ugly. There is an alternative, and that's to use the *vinculum*. Which do you prefer?

$$(((((7 - 5) \times 9) - 6) + 8) \div 4) = 5 \qquad \overline{\overline{\overline{\overline{7 - 5} \times 9 - 6} + 8} \div 4}$$

The brilliant maths communicator James Tanton is on a mission to bring back the vinculum, which admittedly sounds like a medieval torture device but is actually a rarely seen mathematical symbol. It was first used in the fifteenth century but has since fallen out of favour, probably due to the difficulty of printing the symbol with early printing presses. The vinculum works exactly as brackets do, showing the reader which operation should be carried out first.

$$\overline{60 + 60} \times \overline{0 + 1} = 120 \qquad \overline{60 + 60 \times 0} + 1 = 1 \qquad 60 + \overline{60 \times 0 + 1} = 61$$

To its supporters, the vinculum is an elegant way to draw the mathematician's attention naturally towards the required order of operations. To its detractors it looks like a wobbly set of horizontal dominoes that could topple at any point (and it's still quite hard to write it with word-processing software, even some 700 years since the Gutenberg press).

To be fair, there are still remnants of the vinculum in the standard notation that we mathematicians use every day. The sign for a square root is the 'tick'-like symbol that looks like this: $\sqrt{}$. Now, let's say we want to add 9 to 16, and then square root, giving an answer of 5.

$$\sqrt{9} + 16$$

This doesn't look right; written like this it definitely feels like we're being commanded to do the square root first. There's no actual indication in our BODMAS acronym of where to do square roots, but it's implied that the first 'grouping' – square rooting the 9 – should come first, giving a final answer of 19. How about this then?

$$\sqrt{(9 + 16)}$$

This gets the idea across, and you may need to use this layout for some software packages, but it isn't the accepted style. All mathematicians would write the following:

$$\sqrt{9 + 16}$$

where the square root symbol, instead of being followed by brackets, is followed by a vinculum! Not many people realize that the expression above actually contains two symbols: a square root and a vinculum.

Bring back the vinculum? What do you think?

It is quite clear from the construction of this 'sung' arithmetic problem that the usual order of operations should be thrown out in favour of working from left to right: after all, there's no way to add brackets to the spoken (or sung) word. Context is everything. Yet there will still be situations when the written intention of a calculation is unclear, and no amount of BODMAS will get you out of the very deep rabbit hole you find yourself falling into. Hold on tight now...

#27
8 ÷ 2(2 + 2)

One of the undoubted heavyweights of social media maths problems, at the time of writing this, is the number 1 international search return for the phrase 'viral maths', even at a time when the maths of actual viruses is quite prominent in people's minds. Let us take our trusty Swiss army BODMAS rule to the above and then see what happens.

Firstly, there seems to be little doubt that we should tackle the brackets first. Adding the two 2s inside the brackets gives 4, so that we now have:

$$8 ÷ 2(4)$$

Eight divided by 'two lots of four'. Or should that be: 'eight divided by two' lots of four? It seems the problem is asking us to decide whether multiplication or division should take precedence next, but BODMAS does not give a clear steer on this. Hence every time this problem or a similar problem goes viral, the endless resultant comments will be split fairly equally between two camps: the 'multipliers' who multiply first, in this case achieving an answer of 1; and the 'dividers' who divide first, giving a result of 16.

This question is particularly divisive because it seems that calculators can't even decide on the right answer. Entering the phrase '8/2(2+2)' into the Google search engine returns the result 16, with Google reinterpreting your request as '(8/2)*(2+2)'. Most scientific calculators will similarly give an answer of 16, but some models will give the answer 1:

It has also been reported by some sources, such as the British newspaper *The Independent*, that the correct order of operations *changed* in 1917, so that anyone replying to a Facebook thread about this problem who is of the age of approximately 115 or older would actually be within their rights to give a different answer than their younger counterparts. I had heard that the secret to great comedy was… timing, but I had no idea it was also the secret to correct mathematics! Fortunately it seems that reports of changes to BODMAS in 1917 are somewhat exaggerated, and appear to stem from a letter entitled 'Discussions: Relating to the Order of Operations in Algebra' by Professor N. J. Lennes from the University of Montana in a February 1917 issue of *The American Mathematical Monthly* (it's a thrilling read). I won't go into the details here, but Professor Lennes is *very* insistent that multiplication and division should be carried out from left to right if there is no other clear precedence, *and he uses a lot of italics to say so!*

So what is it then, 16 or 1? Just like that blue/black or white/gold dress again, or the Yanny/Laurel sound clip,* it appears there is no end to people's capacity to argue over the binary nature of this problem. One such tweet from summer 2019 that featured this problem gathered tens of thousands of shares and comments, split fairly evenly between 'sixteen-ers' and 'one-ers', often referring to each other with terms that are not exactly kind.

it's 16 omg the replies are embarrassing

It's 1. The amount of people saying 16 need to retake math

i took 3 calc classes, differential equations and linear algebra, it's 16 bro

i have 2 maths degrees it's 1

To settle this once and for all, if we absolutely have to come down on one side or the other, we must understand the following points:

1.   When there is no obvious precedence between multiplication and division, or addition and subtraction, we work from left to right. This was the case before 1917, Professor Lennes argued the case strongly in 1917 and it is still the case now. So the answer to the question is **16**.

---

* Again, please don't look it up now unless you are prepared to lose the rest of the afternoon…

2. *It is a terribly written and intentionally mischievous question.*

Let us be very clear: no mathematician worth their salt would write without clarifying their purpose, or the purpose would be very clear from the context. You should really feel no prouder for answering 16 than answering 1, since the question was intentionally set to confuse you in the first place. This is the mathematical equivalent of walking up to someone in the playground and saying: 'antidisestablishmentarianism is a long word, can you spell it?', and then pointing and laughing in their face if they don't immediately reply: 'IT'. As with so much of what is popular and successful on social media, the quickest way to likes and shares is to cause polarization and division, not to teach good method or correct mathematics.

Look how easy it is to make the question clearer:

If your intention is for an answer of 16: $(8 \div 2) \times (2 + 2)$
If your intention is for an answer of 1: $8 \div (2(2 + 2))$

With just a few carefully placed brackets (or vinculi…) we now make the problem clear and unambiguous, and instead of hurling insults at strangers in foreign countries whom we will never see or meet, we can use our time spent online doing something more useful and fulfilling for the soul, such as learning a new language or educating ourselves in world history or politics. After you…

There can be no doubt that the binary right/wrong nature of this problem is what gives it such appeal, and this can be exploited to create a certain niche type of 'intentional trolling' viral problem, of which the following by maths lecturer and writer Ed Southall is a supremely playful example:

#28

Solve carefully...

230 − 220 × 0.5

You probably won't believe it,
  but the answer is 5!

Credit: Ed Southall

What's going on here then? You're now enough of a BODMAS expert to see that there are two operations, subtraction and multiplication, and the latter takes precedence:

$$230 - 220 \times 0.5$$
$$= 230 - 110$$
$$= 120$$

So why has a respected lecturer and writer taken it upon himself to introduce this bit of intentionally wrong maths into the world? The answer lies in that little exclamation mark at the end: *You probably won't believe it, but the answer is 5!*

    Mathematicians in on the joke will appreciate that the exclamation mark actually has a mathematical meaning, that being the *factorial* operation. To put an exclamation mark on the end of a number means to multiply it by every smaller integer, which will tell you the number of ways of ordering that many items. For example, $3! = 3 \times 2 \times 1 = 6$, so there are 6 ways of ordering three items. If we label the items A, B, C, it's fairly quick to list all the possible orderings, or *permutations*, and see that this is true:

ABC, ACB, BAC, BCA, CAB, CBA

Factorial numbers grow extraordinarily quickly. How long, for example, do you think 10! seconds is? You might like to have a guess before we do any calculations. Oh, but we're not going to use a calculator. Where's the fun in that?

$$10! \text{ seconds} = 10 \times 9 \times 8 \times 7 \times 6 \times 5 \times 4 \times 3 \times 2 \times 1 \text{ seconds}$$

First we might notice that 10×6 is 60 seconds, or a minute, so we can swap out the 10×6 if we switch the units to minutes (there's also no real need to write the '$\times 1$' at the end, since multiplying by 1 has no effect):

$$10! \text{ seconds} = 9 \times 8 \times 7 \times 5 \times 4 \times 3 \times 2 \text{ minutes}$$

If we can find another 60 in there we can switch the minutes to hours, and the most obvious candidate is $5 \times 4 \times 3$, so that:

$$10! \text{ seconds} = 9 \times 8 \times 7 \times 2 \text{ hours}$$

Next we need to cast out 24 to change the hours to days, which at first looks impossible. However, we can exchange the 9 for 3×3:

$$10! \text{ seconds} = 3 \times 3 \times 8 \times 7 \times 2 \text{ hours}$$

Which then allows us to find a 3×8 to exchange for 24 hours:

$$10! \text{ seconds} = 3 \times 7 \times 2 \text{ days}$$

And of course seven days is a week, so that:

$$10! \text{ seconds} = 3 \times 2 \text{ weeks}$$

10! seconds is exactly **six weeks**: no more, no less. That's why maths teachers will write *10! seconds till Christmas!* on their whiteboards on 13 November.

I was first sent this wonderful little gem by email when I started my initial teacher training around 2006. No one could have sent it to me on Facebook or Twitter, *because they hadn't been invented yet*. Well actually Facebook had, but I was too cool (or not cool enough?) to have it by then.

Factorial numbers grow very quickly: 5! is over a hundred (5! = 5 × 4 × 3 × 2 × 1 = 120), but 10! is over a million (10! = 10 × 9 × … × 2 × 1 = 3,628,800).

So when Ed writes 'You probably won't believe it, but the answer is 5!' he means just that: the answer is 5 factorial, which is 120.

Before we move away from factorials (there's a very subtle foreshadowed hint there), and this chapter altogether, here's a fun puzzle to have a go at. Don't forget about your order of operations! Answers at the back for this one…

#29

Make the answer of 6 using exactly three of each digit from 1 to 9. I've done one of them for you. You can use brackets and any operation that doesn't introduce more numerals (so no squaring or cubing).

1   1   1 = 6
2   2   2 = 6
3   3   3 = 6
4   4   4 = 6
5 + 5 ÷ 5 = 6
6   6   6 = 6
7   7   7 = 6
8   8   8 = 6
9   9   9 = 6

Bonus challenge:
0   0   0 = 6

5

BAD MATHS

When Facebook meets algebra

I need to give you fair warning that we're about to head into very some murky waters indeed. As a lifelong lover and advocate of mathematics – especially the kind of ingenious, memorable, mind-expanding nugget that got my student so excited back at the start of the first chapter – very little upsets me more than seeing the internet awash with bad maths.

What exactly is bad maths? I define it as anything that looks like maths, but actually isn't. Mathematics should mean using ingenuity to crack beautifully constructed problems. It should enrich the mind, body and soul. Well, perhaps not the body, but you can't have it all. It absolutely shouldn't be the internet pointing and laughing at you for failing to notice the difference between three bananas and four bananas. It used to be a regularly stated fact – albeit with little actual evidence – that 90% of the content uploaded to the internet was pornography. Well I'm going to state with a similar lack of statistical backing that 90% of the maths on Facebook is bad maths.

Before we get to that though, I think we're going to need to sit down for a good old cup of tea.

#30

A cafe sells tea and coffee. Flynn buys two teas and three coffees and is charged £9.00. Nadiya buys a tea and a coffee and is charged £3.50. What are the individual prices for tea and coffee?[*]

This problem will look very familiar to anyone currently attending British secondary school, and could possibly be solved in the following manner.

If Nadiya doubled her order, logically it would cost her twice as much, so two teas and two coffees will cost £7.00. Note that this is very similar to Flynn's order, with Flynn just ordering one more coffee. Since the difference between Flynn's order and Nadiya's doubled order is £2, this must be the price of a single coffee. Now to find the price of a tea, we simply need to consider Nadiya's original order of one tea and one coffee. Since this combined order comes to £3.50, and the coffee cost £2.00, a tea must cost £1.50.

Again, this feels wordier than necessary, so we might instead use the letters t and c to represent the price of a tea and coffee respectively:

Flynn: $2t + 3c = 9$
Nadiya: $t + c = 3.5$

We then doubled Nadiya's order so that the two equations were more comparable, and followed this by subtracting the second equation from the first:

---

[*] It is an ambition of mine to one day own a cafe that operates this kind of a pricing structure.

'I'd like a cup of tea please.'

'Certainly.'

'Erm… how much will that be?'

'I can't tell you, but I can tell you what the last two customers ordered and how much they paid.'

Flynn: $2t + 3c = 9$
Nadiya: $2t + 2c = 7$
So: $c = 2$

Finally we can substitute into Nadiya's original equation to find t:

$$t + c = 3.5$$
$$t + 2 = 3.5$$
$$t = 1.5$$

We could even substitute into Flynn's equation, just to double check that it all works:

$2 \times 1.5 + 3 \times 2 = 9$, as required.[*]

In mathematics this is known as a set of *simultaneous equations*, where knowing either Flynn or Nadiya's order in isolation would be insufficient to determine the prices of the items, but combining the two is enough to solve the problem. It is worth noting at this point that because there were two unknowns (the prices of tea and coffee) we needed two equations (receipts) to find those unknowns. If there were three unknowns we would need three different equations, and if there were ten unknowns we would need ten equations (though this would be an extremely tedious process that you would certainly want to use a computer for).

---

[*] Note how we have two lots of multiplication and one addition, all together in one line, but we don't need to be told about any 'BODMAS' or 'PEMDAS' rule to know what to do first. It is clear from the context. This is so often the case: outside of school maths papers and social media trick questions there is usually a context that makes the intended order of operations obvious.

Solving simultaneous linear equations is gloriously enjoyable and I remember regularly begging my aforementioned secondary maths teacher to give us simultaneous equations to solve in class. Eventually, after weeks of this, she snapped and said: 'If you enjoy doing them so much, make some up for each other to solve!' So we did.

Problems with simultaneous equations are really easy to create, because you can just think up the solution first and then work backwards. For example, when I made that example with the teas and coffees I decided on the prices first, and then just created a couple of orders that combined teas and coffees in different ways.* If you don't create your question by working backwards it can cause a lot of pain and misery, including for the subjects of your questions. Here's one from a few years ago – original source unknown – that I like to call *When simultaneous equations go wrong*™.

There are 49 dogs signed up to compete in the dog show. There are 36 more small dogs than large dogs signed up to compete. How many small dogs are signed up to compete?

If we let x represent small dogs, and y represent large dogs, we can set this situation up as a set of simultaneous equations, much like we did a couple of chapters ago:

$$x + y = 49 \text{ (49 dogs in total)}$$
$$x - y = 36 \text{ (36 more small than large)}$$

---

* It is important that they are combined in *different* ways: if I told you that 1 tea and 1 coffee costs £3.50, and 2 teas and 2 coffees costs £7, I haven't really given you two different equations, because the second fact is essentially the same as the first. You're still stuck on one equation for two unknowns: not enough to solve the problem.

Adding these two equations together will cancel out the y values, taking us to:

$$2x = 85$$
$$x = 42.5$$

Oh dear. It appears that the dog show features 42.5 small dogs and 6.5 large dogs. Sounds messy. When I said it was hard to go wrong when setting up simultaneous equations questions, perhaps I should have said: difficult, but not impossible.

This ability to quickly create simultaneous equations came in surprisingly handy in my former life as a determined yet failing indie-rock musician, when we had a gig cancelled in Norwich because the venue had no idea we were booked to play and the promoter refused to answer their phone. Actually, the venue did offer us the gig, but for no pay and if we agreed to be finished before 7.45 p.m. when the Champions' League football was due to start, so as not to scare off the locals.

An exciting prospect certainly, but an offer we decided to turn down. Instead we decamped to a different pub where the majority of the band attempted to cover our tour losses by hustling some Norfolk locals at pool, while Jack from our support band convinced me to make up simultaneous equations for him to solve on the back of a beer mat. He hadn't done them since school and missed them, and knowing I was a maths teacher he persuaded me to create him problems to solve for hours on end, while we drank the tour into further debt and wept over our poor life decisions. But the simultaneous equations were at least some consolation.

Try it yourself

1. A venue sells 400 tickets for a show. Adult tickets are £10 and child tickets £8. The venue takes £3900 on the door. How many of each ticket type were sold?

2. A woman walks along a 100m long airport escalator in 25 seconds. She then turns and walks the whole length of the escalator backwards, i.e. 'against the flow', in 50 seconds. What is her walking pace, and how fast is the escalator moving?

3. A mother is 21 years older than her child. In 6 years' time the mother will be exactly 5 times as old as the child. Where's Dad?

From teas and coffees to pencils and pens, here's a classic brain-teaser. It works best if you answer it as quickly as you can.

#31
A pen and a pencil cost £1 in total. The pen costs 90p more than the pencil. How much do they cost individually?

This little gem seems to do strange things to the brain, with many people quickly lurching for an answer of 90p and 10p. Perhaps the mention of 90p in the question tricks the brain to pick this as part of the solution, a bit like saying 'try not to think of a sheep'. (You did it again, didn't you?) The actual answer – **95p and 5p** – does not spring quickly to mind, perhaps because the brain is determined to look for solutions that are nice multiples of ten pence.

If you did take a little longer over the question, you might consider setting it up as a set of simultaneous equations, perhaps using the letters a and b to represent the cost of a pen and a pencil respectively:

$$a + b = 1$$
$$a - b = 0.9$$

This time it's going to be most beneficial to add top to bottom, so that the $+b$ and $-b$ cancel out, leading to:

$$2a = 1.9$$
$$a = 0.95$$

Our pen price could then be substituted back into either of the original equations:

$$a + b = 1$$
$$0.95 + b = 1$$
$$b = 0.05$$

I wouldn't necessarily recommend an algebraic approach for solving this problem; one might even argue that it takes a little of the joy away. I include it merely to show that any algebraic problem like this is solvable

Wonderfully, there is even a maths magic trick based on simultaneous equations. My head nearly fell off when I first saw this.

Write two short algebraic sequences with three terms in each, for example: 3, 5, 7 (going up in 2s) and 10, 7, 4 (going down in 3s). Now turn them into simultaneous equations, for example: $3x + 5y = 7$ and $10x + 7y = 4$.

Now solve your equations. They always have the same solution, *regardless of the numbers you chose!*

(Showing why this works is a fun if fiddly task, so I'll put it at the back of the book.)

if it has the same number of unknowns as unique statements linking them.

If you'd told me aged 13 that there would be a network that connects people around the world, and that it would be accessible at a few seconds' notice from a small device in your pocket, and furthermore that people would use this platform to share simultaneous equations, I'd have been ecstatic. This was more than internet club could ever have promised! But, as is so often the case, the reality is so much worse. Yes, we're about to engage with Facebook fruit equations…

#32

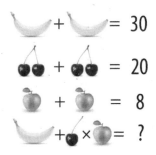

(Source unknown)

Let's first do a few quick checks and balances. We have three unknowns: apple, banana and cherry. There are also three equations linking them, none of which are essentially the same equation, so we should have enough information to find the numerical value of apple, banana and cherry. There is then a fourth line combining the values of a, b and c (there I go slipping in algebra again!), which is probably included as a more satisfying way of a puzzler declaring their solution than giving the three individual numerical values of the fruit.

The mathematics involved is much easier than in the last few examples, so we should be fine. Quickly we can see from the top line that each banana is worth 15, and from the next line that cherries are worth 10 per pair, or 5 each. Similarly apples are worth 4. What do you do next?

If you want to answer '80' turn to page 94.
If you want to answer '35' keep reading.

I know that you wouldn't have fallen into their trap though; from our travels in BODMAS/BIDMAS/PEMDAS we know that multiplication takes precedence – repeat after me: *Potenz vor Punkt vor Strich!* – so that:

$$b + c \times a = 15 + 5 \times 4 = 35$$

What's so bad about this? you might ask. Not much, to be fair; it is essentially another dressed up order of operations question. This is very much an entry level Facebook fruit problem though, as we shall see:

#33

(Source unknown)

This seems fine. The top line makes it very clear that the value of apple is 10, then using this in the next line means the value of bananas must be 4. From line 3 we can then quickly see that coconut is worth 2, so that coconut + apple + bananas = 16. Easy! Easy, but WRONG!

What did we do wrong? Look more closely... the banana images in lines 2 and 3 are bunches of four bananas, but the bunch in the bottom line is only three bananas. It's actually the case that each individual banana is worth 1, so that the final sum is coconut + apple + three bananas = 2 + 10 + 3 = 15. And that's what I call bad maths. This is not a maths problem, it's a point-and-laugh-at-people-who-don't-spot-the-4th-banana problem. Unfortunately, we're still only scratching the surface:

#34

(Source unknown)

This time I'll give you some options:

A) 135   B) 51   C) 39   D) 38

Here we go. There are some bananas so we can probably assume that there will be a sneaky extra banana somewhere, and it seems there is: the bunches in rows 2 and 3 have four bananas per bunch, but the bunches in the bottom row have three. Also we have some inevitable BODMAS action going on in the bottom row, but we're far too experienced by this stage to fall for that.

So from the top row we can see that the shapey thing is worth 15, and the second row shows that a four-bunch of bananas is worth 4, so each banana is worth 1. All good so far. Four bananas plus two clocks is 10, so each clock is worth 3. So to finish off (using the letter a to represent the shapey thing):

$$c + 3b + 3b \times a = 3 + 3 + 3 \times 15 = 3 + 3 + 45 = 51$$

WRONG! Oh dear. You might have spotted that the shapey thing is also not the same in the bottom row as in the other rows. It's… missing something. If you look very carefully you'll see the first shapey thing is a square within a pentagon within a hexagon, but the bottom shapey thing is simply a pentagon within a hexagon. If you're starting to lose patience with this whole problem I absolutely empathize, but please just persevere for a little longer.

The square–pentagon–hexagon shapey thing is worth 15, so is there something inherently 15-ish about that shape? There is indeed: a square, pentagon and hexagon combined have 15 sides, so it appears the rule for any compound shape is to add together all the sides. The shapey thing at the bottom has 11 combined sides, so:

$$3 + 3 + 3 \times 11 = 3 + 3 + 33 = 39$$

WRONG! Give me strength, what now? The *very* observant may have noticed that not all the clocks are the same either. The first two clocks are set to 3 o'clock, but the clock in the bottom row is set to 2 o'clock. The rule for clocks, it seems, is that we assign it the value of the hour the clock is set to. One more time then:

$$2 + 3 + 3 \times 11 = 2 + 3 + 33 = 38$$

Which I think is the intended solution, but honestly, who cares? I'd actually be ashamed to get this question right first time. Where's the joy, the intrigue, the spark? Every time my auntie tags me into one of these threads and I see thousands of shares and comments, a little bit of my mathematical soul dies. What are we actually aiming for here? A nation of people who are adept at basic arithmetic and observing from a glance the difference between a clock set to 2 o'clock or 3 o'clock? In the coming years will I be checking UCAS applications from students boasting their skills in algebra, calculus and quickly being able to ascertain the difference between three or four emoji bananas?

For me, this question represents the nadir of bad maths (and I have suffered many of these questions in researching this book) in part because it takes a shape you can do some really nice maths with – a square inside a pentagon inside a hexagon – and instead uses it for sneaky trolling.

Though I hate to dwell on this point for any longer than I need to, the clock faces particularly upset me. Since finding the previous problem I've done further research and uncovered a whole sub-genre of clock-based bad maths problems, such as the following abomination:

#35

(Source unknown)

The naive puzzler looks at line 1 and sees that a clock is worth 7... WRONG! The clocks in the top line are not all identical. After some careful (if grudging) consideration, it looks like two clocks showing 9 o'clock and one showing 3 o'clock, so that 9 + 9 + 3 = 21. But... these clocks have eight hour positions, not twelve! The clocks don't actually show 3 and 9 o'clock, but rather 2 and 6 o'clock, in base eight. A few countries have attempted to decimalize time – that is, split the day into ten units rather than 24 – most notably France during the revolution in the late eighteenth century,* but to my knowledge no culture or civilization has ever attempted to convert time into base eight. What are we to make of this nonsense?

I can barely be bothered to complete the rest of the problem. I imagine the calculators, though appearing to be identical, are actually cartoon representations of the classic Canon AS-120 and Casio CA-53 models, and these numbers must somehow be substituted into the

---

* If you're wondering, the day was split into ten decimal hours, with each of these broken down into 100 decimal minutes, each containing 100 decimal seconds. Somehow, it never caught on.

equations. Again, who could honestly care?*

Sometimes genuine attempts at writing a decent emoji equation can lead to some amusing and unexpectedly challenging mathematics (in a good way):

#36

Credit: Twitter/CBeebiesHQ

This problem was shared on the parent-facing social media channels of CBeebies, the BBC's pre-school TV channel, to promote the show *Apple Tree House*. First and foremost let me state my position: I am a huge fan of CBeebies. It has so many educationally enriching shows: *Alphablocks* and *Numberblocks*, which teach kids phonics and (surprisingly subtle) mathematics; *Go Jetters*, which teaches worldliness and advocates cultural understanding; *Bing*; and *Hey Duggee*, which promotes teamwork and lifelong learning. I have watched hundreds of hours of CBeebies with my children and we all love it, so CBeebies would have to do a lot to upset me.

Having said that… let's break this question down. First and foremost, there doesn't seem to be any emoji-based subterfuge going on: the trees

---

* Oh all right, if you must. I *think* the intended answer is 414. Top line: 9 + 9 + 3 = 21. Next line: each calculator shows 1234, which adds up to 10. 10 + 10 + 10 = 30. Third line: Lightbulb = 15. Bottom line: The calculator says 1224 not 1234! So 9 + 9 × 45 = 414. Ugh, I need to go and take a shower now.

are trees, the houses are houses. Amen to that. We also have to make a careful decision with the bottom row as it contains multiplication and subtraction, but the multiplication is on the left-hand side of the subtraction anyway, so that shouldn't cause any problems, whether people know about BODMAS rules or not. In fact, this whole thing looks far from sinister, one might even say innocent.

Perhaps a little too innocent. There is one major way in which this diverts from the usual emoji-maths template: there are only three lines, the top two of which are equations. Three unknowns (apple, house, tree) but only two equations; that's not enough information to solve the puzzle. Or rather, not enough information to find a *unique* solution. Rather, we will find a whole family of solutions. A popular response in the comments was that apple, house and tree represent 1, 10 and 15 respectively, giving a final answer of 14. But one poster argued that apple was worth 13, house 2 and tree 11, giving an answer of 130, which seems to work equally well. Which is it, both? And if both, how many more possible solutions? May I take us, for a moment, back to the teas and coffees example?

A cafe sells tea and coffee. Flynn buys two teas and three coffees and is charged £9.00. Nadiya buys a tea and a coffee and is charged £3.50. What are the individual prices for tea and coffee?

Flynn: $2t + 3c = 9$
Nadiya: $t + c = 3.5$

If Flynn and Nadiya were attempting to solve this problem themselves, they might try to think of all the possible prices that a tea and a coffee could have. For example, Nadiya might guess from her order that tea costs £1 and coffee £2.50, or perhaps tea costs £2 and coffee £1.50, or she might get tired of writing all this out in a big long sentence and simplify with a table.

Tea cost	Coffee cost	Total cost ($t + c$)
£1.00	£2.50	£3.50
£2.00	£1.50	£3.50
£3.00	£0.50	£3.50

Flynn might do something similar from looking at his receipt:

Tea cost	Coffee cost	Total cost ($2t + 3c$)
£0.75	£2.50	£9.00
£1.50	£2.00	£9.00
£2.25	£1.50	£9.00

But still neither of them would know the price of a single drink without comparing each other's receipts. If both Nadiya and Flynn plotted the points from their tables on a graph, the result would be as follows.

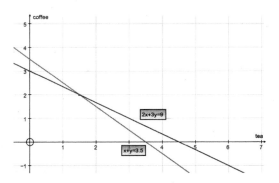

Here one graph represents all the possible tea/coffee prices that work for Nadiya's receipts, and the other graph represents the same for Flynn. There's only one point that lies on both lines, and that is the solution we seek: £1.50 for a tea and £2.00 for a coffee. Now back to CBeebies.

Credit: Twitter/CBeebiesHQ

They give two equations that link three unknowns, so we need an axis each for apple, tree and house. Because we are now in three dimensions not two, the sets of points that satisfy each CBeebies equation will be a plane instead of a line – that is a giant 3D sheet of solutions, like an infinite flat sheet of paper, stretching endlessly in all directions. Because the problem only specified two equations, we are interested in where two planes intersect. To imagine how two planes intersect, simply look at the book you're holding right now.* The left-hand page and the right-hand page are both planes, and they intersect along a line – that is, the spine of the book. The spine, by definition, is the line along which the two book pages meet. Because one page represents

---

* Kindle readers – I apologize for excluding you. Please either imagine a real-life book or fold your Kindle in half. If you're on the Tube, find a copy of the *London Evening Standard*; there's always one lying around. (Non-Londoners – I apologize for excluding you.)

the first CBeebies equation, and the other page represents the second, there must be an infinitely long line of points that represent values for apple, tree and house that work for both equations.

As it happens, tree can be any value you like. Then house will be double your tree value minus 20, and apple will be 46 minus triple your tree value (workings at the back of the book):

$$(a, h, t) = (46 - 3t, 2t - 20, t)$$

So, for example, if tree is worth 15, then apple is worth 1 and house is worth 10, giving a final answer of 14. If tree is worth 10, then apple is 16 and house is 0, giving an answer of 144. I could continue this paragraph indefinitely since there is an infinite set of values that satisfy the top two equations, but the publisher says this is not an acceptable way to hit my word count.

This means that the bottom line, apple × tree – apple, can be practically whatever you want it to be. I say practically, because in actuality you can't make any values above 154.083, but you can make anything else. I'm sorry if you were craving one particular solution to this problem, but the good news is you could essentially have come up with **anything less than 154.083** and justifiably claim a correct answer. Again, the maths behind all this is at the back of the book, but it did keep my A-level further maths class busy for a good half-hour or so, which was certainly a more enriching mathematical experience than all of the three-or-four-bananas emoji questions in the world combined. At least bad maths isn't being explicitly taught in schools yet…

#37

$$\left(\ddot\smile\right) + \left(\ddot\smile\right) + \left(\ddot\smile\right) = 27$$

$$\left(\overset{\star\star}{\smile}\right) + \left(\ddot\smile\right) \times \left(\overset{\star\star}{\smile}\right) = 80$$

$$\left(\overset{\star\star}{\smile}\right) + \left(\overset{\star\star}{\smile}\right) \times \left(\overset{\frown}{\frown}\right) = 48$$

$$\left(\overset{\star\star}{\smile}\right) + \left(\overset{\frown}{\frown}\right) \times \left(\ddot\smile\right) = ?$$

Here we observe many of the hallmarks of bad maths, including some sneaky hidden emojis in the bottom line. I believe the intended answer is **106**, if somehow you're still interested.[*] What I find fascinating about this is that it was set as homework for a ten-year-old as they transitioned from year 5 to 6 at primary school. While some people enjoy these problems very much – you may have noticed that I'm not one of these people – this strikes me as an extremely contentious piece of work to set children on a long break at home over the summer holiday.

Firstly, while the teaching of order of operations does take place at primary age, it's a very difficult concept for ten-year-olds to grasp, with nuances that I have discussed briefly in the last chapter but which parents less confident with maths could definitely struggle with. Setting this as work over a six-week summer break seems extremely risky.

Secondly, the 'gotcha' aspect of the hidden emojis in the bottom line does not really represent the message we would want young people to take away from mathematics. There is a beauty in logical deduction and

---

[*] Top line: smiley = 9. Next line: star-face plus nine star faces = 80, so essentially ten star-faces = 80, therefore star-face = 8. Third line: 8 + 8 lots of frowny = 48, so 8 lots of frowny = 40, so frowny = 5. Bottom line: two star-faces = 16, frowny = 5, two smileys = 18. $16 + 5 \times 18 = 106$.

mathematical abstraction that is entirely missing in these questions; I'd go so far as to say these emoji bad-maths questions, at their worst, are flipping the bird in the face of mathematics. Teaching mathematics *via* these questions feels like training police recruits by showing episodes of *Brooklyn Nine-Nine*. I found the setting of this question to a young maths class even more galling when I saw it was accompanied by the following puzzle, with no other supporting information:

#38

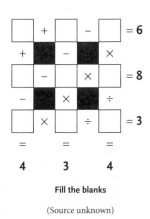

Fill the blanks

(Source unknown)

These problems are reasonably popular and can often be found on the puzzle pages of several major newspapers. Unfortunately, you need to be fairly familiar with these problems to know that there are a few pieces of crucial information missing. For one, these puzzles have nine spaces to fill in because they usually require each of the digits 1–9 to be used once and once only. Secondly, it's really important we are told whether to carry out calculations from left to right (or top to bottom) or if instead we should follow the BODMAS order of operations rules. Neither is better than the other in this context – it's a puzzle and is absolutely allowed to make up its own rules – but it is important that we know what those rules are.

I have to assume that this has been copied from a newspaper or puzzle book, and therefore we must assume the rule requiring single use only of digits 1–9. If this is the case then we can't possibly complete the puzzle if we're keeping to standard BODMAS order of operations rules: the middle horizontal row demands a 9 in the leftmost position, as there would be no other way to 'get down' to 8 for the final row answer. Even if we do have a 9 in the leftmost position, the only option that remains is to have 1 in the two other positions, giving $9 - 1 \times 1 = 8$, which works but breaks the single-use rule. We therefore can only guess that we throw out the BODMAS rule and work from left to right or top to bottom.

After ten minutes or so of playing with this I was able to find an arrangement of nine digits that works, but this was having ascertained the rules of play for myself – see my solution at the back of the book. I'm not sure the average ten-year-old, or their parent, would have the patience to do this, nor should they have to! I suppose what I'm saying again is that context is everything. This is a great puzzle as an optional exercise with very clearly laid out rules, but as a compulsory task with missing instructions – and instructions that, when realized, contradict those of the previous puzzle! – it is likely to frustrate and antagonize students and their parents, potentially giving a lasting bad impression of what mathematics is and isn't.

The emoji-equation template we've seen throughout this chapter seems to be such a well-established trope that it can be quite jarring, though refreshing, to see an emoji puzzle playing by different rules.

#39

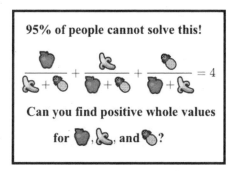

Credit: r/MathWithFruits

Before you spend too long on this I'd definitely recommend widening the rules somewhat to allow for negative numbers. Even finding a solution that includes negative numbers is tricky, but finding a solution that uses only positive whole numbers... well we'll come to that shortly. But let's just say for now that it is profoundly difficult.

The problem can be solved by trial and error once we allow for at least one of the fruits to represent a negative number, and I have to confess this problem swallowed up a good few hours of my time even with that adjustment. At first I tried to use a set of numbers that would create three 'similar' denominators (that's the mathematical name for the bottom of a fraction) but after some concerted effort that approach proved to be fruitless (I'm sorry...).

I then created a spreadsheet that would randomly generate three whole numbers (integers) between −20 and 20 and run them through the equation, hoping to find an answer of 4. After running 700 or so trials (which took a matter of seconds) I was able to find a solution set, namely **4, 11 and −1**. I was as pleased as Punch (I'm so sorry...).

If I had tried to use my spreadsheet to find a solution set that only uses positive integers I would have been plugging away for a little longer. In fact, the smallest positive integer solution set requires three positive integers that are more than 80 digits long:

apple = 154476802108746166441951315019919837485664325669565431700026634898253202035277999

banana = 368751317941299998271978115652254748254929799689719709962831374716372246340555579

pineapple = 4373612677928697257861252602371390152816537558161613618621437993378423467772036

No computer package could find those values by what we would call 'brute force', i.e. running through every set of three integers until it landed upon that set above.* There simply isn't enough computing power (or time) in the world. So how did anyone find them?

Because this problem requires a whole-number solution, it falls under an area of maths known as number theory. Surely maths only involving whole numbers must be much easier, right? That's infinitely many nasty fractions and decimals that we don't have to worry about! Unfortunately I can confirm from my degree in mathematics that number theory ranges from 'pretty difficult' to 'cripplingly, mind-meltingly, day-spoilingly difficult', and this problem is certainly closer to the latter.

I am not going to go anywhere near showing how to solve the

---

* Although remarkably you can, if you're very patient, type those values into a standard spreadsheet package and it will merrily turn the levers and report back the correct answer of 4.

originally stated problem: I couldn't if I tried. All I'll say here is that if your first instinct was to multiply through by all the denominators, that was the right way to go.

$$\frac{a}{b+c} + \frac{b}{a+c} + \frac{c}{a+b} = 4$$

$$a(a+c)(a+b) + b(b+c)(a+b) + c(b+c)(a+c) = 4(b+c)(a+c)(a+b)$$

Unfortunately, at this point you are faced with a third-order diophantine equation, in other words a whole world of pain. If you are interested in going further, the mathematician Alon Amit has written a brilliant summary online.[*] Amit also expands a little on the accuracy of the '95%' clickbait headline. Are 5% of the general public conversant in third-order diophantine equations? Of course they aren't. To quote Amit directly, 'Roughly 99.999995% of the people don't stand a chance at solving it, and that includes a good number of mathematicians at leading universities who just don't happen to be number theorists. It is solvable, yes, but it's really, genuinely hard.'

FACT CHECK!

I really can't help myself from unpacking this a little. The most recent Times Higher Education university rankings lists 1400 reputable universities, each of which might have, say, five number theorists who could reasonably have a good crack at that problem. If so, that's 7000 people out of 7 billion (very roughly!), which makes it quite literally a one-in-a-million problem, in other words 99.9999% of people can't solve

---

[*] www.quora.com/How-do-you-find-the-positive-integer-solutions-to-frac-x-y+z-+-frac-y-z+x-+-frac-z-x+y-4. I highly recommend this article but you need quite a firm grounding in advanced mathematics.

this problem. Amit seems to have assumed even fewer mathematicians in the world who would be adept enough in number theory to solve this, but either way I think we can certainly say it's markedly fewer than 5% of the population. As one Reddit user put it, '... saying 95% of people can't solve this is like saying 95% of people can't jump over a skyscraper'.

In fact, this problem seems to have originated from the 'Math with Fruits' Reddit page, in which genuinely very hard maths problems are made apparently Facebook-friendly by turning variables into fruit emojis and pasting in a made-up clickbait percentage, usually '95% of people can't solve this!' It appears this particular problem originates from a 2014 paper by the Scottish mathematician Allan MacLeod about some profoundly hard maths, and some scamp swapped the variables for fruit and got it to a much wider audience. Here's another gem from the 'Math with Fruit' page:

#40

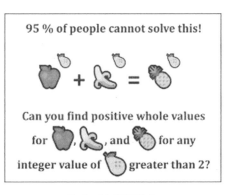

95 % of people cannot solve this!

Can you find positive whole values for 🍎, 🍌, and 🍍 for any integer value of 🥝 greater than 2?

Credit: r/MathWithFruits

Again, I'll unpack a little for those less familiar with this kind of problem. If we let the lime equal 2, we are essentially looking for values of x, y and z that would satisfy:

$$x^2 + y^2 = z^2$$

This you may recognize from your own mathematics education as Pythagoras' theorem, and it does indeed have some whole number solutions:

$3^2 + 4^2 = 5^2$ $(9 + 16 = 25)$
$5^2 + 12^2 = 13^2$ $(25 + 144 = 169)$
$7^2 + 24^2 = 25^2$ $(49 + 576 = 625)$

Actually it has more solutions, but I won't list them all here. Are there any if we replace the lime with 3 instead of 2 though?

$$x^3 + y^3 = z^3$$

Here's one that's close:

$6^3 + 8^3 \simeq 9^3$ $(216 + 512 = 728$, one short of 729$)$

But I wouldn't waste time looking for a better fit; that's as close as you're ever going to get. It's also impossible to find a set of x, y, z for which the condition would hold if the lime represented 4, 5, 6 or any other larger integer. Herein lies the joke: this was for many years one of the most infamous problems in all of mathematics, known as Fermat's Last Theorem.[*]

---

[*] If Fermat's Last Theorem is new to you I highly recommend Simon Singh's fantastic and very readable book of the same name.

Mathematicians suspected that the condition could not be met for any power larger than 2, and this was originally stated without proof by Pierre de Fermat in 1637. It was eventually proved by Andrew Wiles in 1995, by which time around 20 billion people had been born and died since Fermat's original conjecture.[*] Wiles did have some significant assistance from Richard Taylor in completing the proof, so perhaps we might say that 2 people out of 20 billion were able to solve this problem. Or, to use the standard Facebook fruit parlance: 99.99999999% of people can't solve this!

We're going to move away from fruit emojis now, if only for the sake of my own sanity.

#41

$$2 + 3 = 10$$
$$8 + 4 = 96$$
$$7 + 2 = 63$$
$$6 + 5 = 66$$
$$9 + 5 = ???$$

(Source unknown)

Not that this genre of puzzle doesn't have its own downsides, but I certainly find these problems less problematic than the previous type.

As a teacher, I've often used this kind of game with students when introducing the idea of a *function*. A function will return an output for any input value (or values): for example, in the above case, when the input values are 2 and 3, the output is 10, and so on. In class I would

---

[*] https://www.prb.org/howmanypeoplehaveeverlivedonearth/

ask students to fire input values at me and I would reply with the output value, until they had tested enough values to guess what 'rule' I was applying to their values.

How to go about solving the above problem? Personally, I noticed that the output values are all *composite* numbers, that is, they can all be written as the product of two other numbers. 10 is the product of 2 and 5, which may relate to the 2 and 3 it came from. 96 can be written in many different ways as a product of two values, so I'll leave that for now. 63 is 7 × 9, and 66 is 6 × 11, and at this stage I noticed that one of the product numbers is part of the left-hand 'sum'. With a little more thought we can work out the final rule – multiply the first number by the sum of the two numbers. Algebraically this could be written as:

$$f(x,y) = x(x + y)$$

So the answer to the 5th line is '9 + 5' = 9(9 + 5) = **126**. Interestingly, this function is not 'symmetrical', so that '5 + 9' = 5(5 + 9) = 70, not the same.

I do like these puzzles, but I find it infuriating that they seem to persist in using an addition symbol as the operator. There are so many symbols available, here are just a few: ⇄ ⚒ ↢ ♬ 😹. Why reuse a symbol that already has a very clear and well-known use? Of course I've answered my own question: the addition symbol is being used for something that very clearly isn't addition, acting as 'clickbait' to tempt the potential puzzler. Again, this is fine for adults who've learnt what all the mathematical operators mean and what they're used for, but we've seen that these 'viral' problems can sometimes find their way into children's hands (or even homework), where they could give a very misleading idea of how mathematical symbols should and should not be used.

#42

$$1 + 4 = 5$$
$$2 + 5 = 12$$
$$3 + 6 = 21$$
$$8 + 11 = ??$$

(Source unknown)

This example goes even further in 'trolling' the puzzler, in that the first line works for 'normal' addition, but the other lines don't. But I still don't see that there's any advantage in using the addition symbol for this kind of puzzle, as certainly it would work equally well with any other symbol in its place. (The rule here, by the way, is 'multiply the first number by the second number plus 1', or $f(x,y) = x(y + 1)$, giving a final answer of **96**.)

I'm aware that this chapter may have come across somewhat negative, and perhaps I'm wrong to have anything other than the very lowest expectations of the ~~moral cesspit~~ established social media platform that is Facebook. To attempt to brighten the mood, here are some other problems involving number patterns and connections that I've seen spread widely online in many different guises, all of which I hope you'll agree feature a more satisfying denouement.

#43

What comes next in this sequence?

  1, 11, 21, 1211, 111221, ...

This is known as the 'look-and-say' sequence, and if you haven't cracked it yet, the name of the sequence gives you a clue. If you need a further clue, say the digits out loud – in English.[*]

The next term would be **312211**, because each term describes the previous term. So 1 is followed by 'one one', which we write as 11. This is 'two ones', so next we write 21. The fifth term, 111221, is 'three ones, two twos and one one', so it is followed by 312211, and then 13112221. This sequence grows indefinitely, forming longer and longer strings of numbers, but you will never find any digit other than 1, 2 or 3 in any term in the sequence. Can you work out why? The answer to that one is at the back of the book.

#44

You might need a pencil and paper for this one, or you might be able to do it in your head. It depends how old you are...

- Write down your age.
- If your age is even, halve it. If it's odd, triple it and add 1.
- You now have a new number. Again, if it's even, halve it. If odd, triple and add 1.
- Keep going until you decide to stop.

I often use this as a warm-up when I'm performing family shows, to show the stark difference in outcome for audience members of very

---

[*] I apologize to any non-English readers: I don't in any way mean to assert that my own language is superior to others, but this sequence is too lovely not to include.

similar ages. For example, an eight-year-old would follow this path:

$8 \to 4 \to 2 \to 1 \to 4 \to 2 \to 1 \to$

And this will go on forever or until you get fed up. Again, not a satisfactory way to meet my word count, apparently. But a seven-year-old would encounter this sequence:

$7 \to 22 \to 11 \to 34 \to 17 \to 52 \to 26 \to 13 \to 40 \to 20 \to 10 \to 5 \to 16 \to 8 \to$
$4 \to 2 \to 1 \to 4 \to 2 \to 1 \to$

In other words a seven-year-old has to go through 13 iterations before reaching the starting point for an eight-year-old. Luckily there aren't too many 27-year-olds at my family shows, as they would have to grind through 111 steps, at times carrying out calculations in the 7000s, before eventually settling at the 4, 2, 1 point.

This sequence is variously known as hailstone numbers, wondrous numbers or – my personal favourite – HOTPO numbers ('half or triple-plus-one'). It is thought that all numbers will eventually settle at the 4, 2, 1 point, but although no number has ever been found that doesn't eventually settle, it remains to be proven that every number will get there. For this reason, this is known as the Collatz *conjecture*, named after Lothar Collatz, a conjecture being a conclusion or a proposition which is suspected to be true, but for which no proof or disproof has yet been found.

I've long enjoyed having fun with these numbers, and at one point decided it would be a good idea to set a Collatz chain off on Twitter, hoping to track all 111 steps as they made their way around the world tweet by tweet, like a message in a bottle cascading off waves before settling on a distant shore.

Kyle D Evans
@kyledevans

n = 27

If n is even, quote this tweet and replace n with n/2
If n is odd, quote this tweet and replace n with 3n+1

Unfortunately the outcome I feared most was exactly what happened; rather than one neat linear stream of tweets, there were instead dozens of competing threads, spiralling around maths Twitter and getting firmly on everyone's nerves. I intended to pop a message in a bottle and wave it off to the horizon; what I actually did was set off a hundred identical messages on balloons that got instantly stuck in local trees and suffocated the wildlife. I believe the longest sustained chain was around 33 steps long and fizzled out deep in Spanish maths Twitter on *n* = 263, where, instead of retweeting, someone decided to post a link to a Phil Collins video. The cruellest of epitaphs.

Here's one last problem you can again do either with pencil and paper or in your head, but you must go as quickly as you can.

#45

- **Start with 1000**
- **Add 40**
- **Add 1000**
- **Add 30**
- **Add 1000**
- **Add 20**
- **Add 1000**
- **Add 10**
- **Say your answer out loud as you turn the page**

4100!

Did you say 5000? Don't worry, lots of people do (I did). This cleverly constructed little gem is in no way unique or new, but does work surprisingly well and regularly. By creeping up the thousands and the tens alternately, it seems to trick people to want to tick over the thousands instead of the hundreds when the tens can't go any further. We see the 'gotcha' element of viral maths once again, which made this a big hit on TikTok in the summer of 2020.

I can't help but notice the similarity to the 'carrots' problem we saw back in the playground chapter, and feel somewhat reassured that, however young people are communicating in another 20 years' time – VR helmet? Telepathy? Google brain? – they'll probably be using it to joyously share the very same maths they spend most of their lesson time complaining about.

6

GET INTO SHAPE

Genius geometry problems

#46

What proportion of the square is shaded?

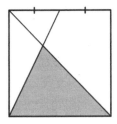

Credit: Ed Southall

How do you feel when you see a problem like that? Excitement or dread? Does it make you want to rush to find a pencil, or lob the book out of the window? You might be expecting me to say that I'm firmly in the former camp, but – shock announcement incoming – I'm more like the latter. This kind of question can't help but remind me of the 'magic eye' problems that became hugely popular in the 1990s, in which staring at a grainy, blurred two-dimensional pattern would miraculously reveal a hidden 3D object. Apparently. I've never been able to see a magic eye illusion, and to this day I remain fairly sceptical that the whole thing wasn't a *Truman Show*-esque joke at my expense being played by the whole world. I sometimes feel similarly about questions like the above: some people can apparently glance at it and know exactly what to do straight away, and that just makes me feel less willing to dig in and persevere. This chapter will act as my attempt to improve myself, and perhaps to help you do the same.

Geometry questions act as my way of empathizing with people who can't stand maths or think they're 'not a maths person'. In truth, of course, anyone *could* be a maths person, if they'd just found a spark and persevered a little more doggedly at a point in their youth when it became difficult. I don't blame people at all: maths certainly does become difficult, and it takes great resourcefulness to power through the hardest parts. Without a certain kind of support it's no surprise that many people lose their way. This is exactly where I feel I am with geometry problems; I feel like I *could* be good at them, but I decided at about 15 that they were my least favourite part of maths, and ever since then I've avoided them, thinking of myself as 'not a geometry person'. There's definitely a 'fight or flight' aspect to any maths question, where it's easier to reject a question as 'too hard' or 'not my cup of tea' than to take the necessary, and often ultimately satisfying, steps to break it down and solve it.

Geometry questions have become increasingly popular on social media in recent years, thanks in no small part to the sterling work of some ingenious puzzle setters who have been kind enough to talk to me about how they do it. If you already feel pretty confident about geometry, I hope this chapter will still present you with some really smart problems that you might not have seen before. Oh, and the answer to the triangle puzzle is $\frac{1}{3}$, but the more important *why* is coming up. Let us begin:

CIRCLES

#47

Which would you rather have, an 18-inch pizza or two 12-inch pizzas?

Before we tackle this, let's rewind a little. What is a circle? Of course I know you know what a circle is, but how would you explain it to an extraterrestrial visitor? How would you explain it to your four-year-old niece? You might say it's the simplest two-dimensional (2D) shape: the shape with only one side and no corners. But a circle isn't the only shape with one side and no corners; ellipses have this feature too.

Besides, if it's the 'simplest' 2D shape, it should look like the next simplest, that being the three-sided triangle. But actually a circle looks an awful lot more like a 15-sided pentadecagon than it does a triangle, and even more like a 50-sided pentacontagon:

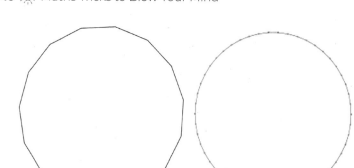

Pentadecagon: 15 sides   *Pentacontagon: 50 sides*

So perhaps a circle has not one side, but infinitely many infinitesimally short sides. I feel like we might be losing our four-year-old niece at this point though. Shall we instead say it's the set of all points that are equidistant from some central point. So if you had a horse tethered to a pole, the tracks that marked the furthest it could move from said pole would form a circle.

If you haven't done anything with circles since school you will hopefully still remember the following symbol: π, which we call 'pi'. If you measure the diameter of a circle (all the way across from one side to the other, through the centre) you will find that almost exactly three of these would fit around the circumference of the circle. I say 'almost' three, because it's actually pi, which is more like 3.141592653589793… (it goes on endlessly and never repeats, but NASA use pi to 15 decimal places, so that's good enough for me).

So the circumference of a circle is π × diameter, or 2 × π × radius, but what about its area? If you cut a pizza* into as many very small slices as

---

* Mathematicians are obsessed with pizza and pies. Actually, in Portugal a pie chart is known as a pizza chart. In France a pie chart is known as a Camembert, but now I'm going off the subject (and getting hungry).

you can (which is a great idea by the way – it means you can have more slices for the same number of calories) you could rearrange them into something that looks a bit like a rectangle.

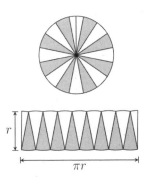

Credit: wikipedia/Jim.belk

The rectangle is of course made up of the same pieces as the original circle, so their areas must be the same. The height of the rectangle is very close to the radius of the original circle, and the thinner you take your slices the closer it will be. Likewise, the width of the rectangle is very close to half of the 'crust', i.e. half of the circumference of the circle, which we established above is half of $\pi \times$ diameter, i.e. $\pi \times$ radius. So the area of a circle is very close to πr^2 or pi $\times$ radius $\times$ radius, and will get closer and closer to this limit as we cut our pizza into thinner and thinner slices.

We can finally return to the original problem now: which is more pizza, an 18-inch pizza or two 12-inch pizzas? It certainly feels like the latter should be correct, but if we trust the maths:

18-inch (9-inch radius)   Area = $\pi \times 9^2 \simeq 254$ square inches

$2 \times$ 12-inch (6-inch radius)   Area = $2 \times \pi \times 6^2 \simeq 226$ square inches

The **18-inch** pizza is larger. Mad, isn't it? The reason this feels so counterintuitive is that our brains tend to think mostly in terms of linear proportion, so that as one variable increases, we expect other variables to increase or decrease at a steady rate. As a case in point, I once heard the radio presenter Mike Parry arguing vociferously that, one day, a man would run the 100m in under 1 second.* His argument was that the world record time is going down all the time, so follow that trend forward far enough and you'll have men running at whatever speed you like. Of course he has overlooked the fact that, although the world record time is decreasing, the rate that it's decreasing at is slowing, and is also currently nowhere near 1 second.

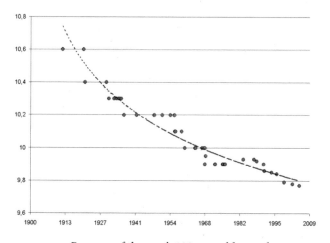

Progress of the men's 100m world record

Credit: wikipedia/mrnett1974

Fantastically, Parry has several times backed up this claim on social media, including – in one of my favourite tweets of all time in October

---

* Yes, it's my own fault for listening to sport radio. It was a moment of weakness.

2012 – that 'Of course the 100m will one day be run in 1 sec but probably not in our lifetime.'

What I like most about this is the 'probably not in our lifetime'. Mike Parry is 66 years old at time of writing. The men's 100m world record has improved by roughly 1 second in 100 years, and the rate of improvement is drastically slowing. I think you are probably safe with that 'probably', Mike.

(Of course, Mike has made many mistakes here. Firstly: not all relationships are linear. Secondly: real-life mathematical relationships – whether linear or otherwise – do not hold forever. Just because the 100m record has dropped regularly over the last 100 years, that doesn't mean it will continue to do so. Most experts agree that human physiology limits a person to sprinting at around 13 metres per second at top speed,* limiting the potential best 100m record at just over 9 seconds. If a person could run 100m in 1 second, it would require accelerating out of the blocks at $20g$, or 20 times normal Earth gravity. When we consider that elite trainee astronauts usually lose consciousness at around $12–14g$ in a centrifuge, this looks unlikely. Mr Parry would do well to remember that there are only two types of people in the world: those who extrapolate unreasonably from small data sets.)

Anyway, not all relationships are linear, and just as the 100m world record is being improved at a slower rate as time passes, the area of a circle increases at a faster rate as the radius increases. The area, being πr^2, means that area is proportional not to the radius but the *square* of the radius, so that increasing the radius by a bit increases the area by a lot, leading to the surprising result that one 18-inch pizza is a better choice than two 12-inch pizzas. Well, unless you prefer the crust to the non-crust, but let's not go down that road.

---

* https://www.wired.com/2008/08/bolt-is-freaky/

While we're on the topic of pizza... what's the volume of a pizza with radius z and thickness a?

If we can find the area of a circle, we can convert that into the volume of a cylinder by multiplying by the depth, so that:

Volume $= \pi z^2 \times a$

$\qquad = \pi zza$

$\qquad = (\text{pi})zza$

TRIANGLES

Now that we've learnt a little about circles, let's consider the humble triangle. There are some crucial things you'll want to know about triangles:

Firstly, the area of a triangle is $\frac{1}{2} \times$ base $\times$ height, as long as the base and height are *perpendicular*, i.e. at right angles. This formula simply stems from the fact that a triangle fills exactly half the space that a rectangle with the same dimensions would.

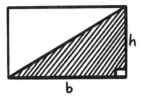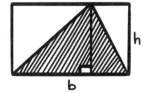

Next, if we know two lengths on a right-angled triangle, we can find the third using Pythagoras' theorem: the square of the hypotenuse (longest side) is equal to the sum of the squares of the two shorter sides.[*] You'll probably remember this from school – or at least from one of the fruit emoji problems in the last chapter – and if you do a stock image search for the word 'mathematics', you'll almost certainly find an image something like this:

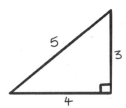

Squaring the short sides here gives $3^2 + 4^2 = 9 + 16 = 25$, so that the longest side can only be $\sqrt{25}$, i.e. 5.

---

[*] Not, as Homer Simpson declares when he wears a pair of glasses he found in the toilet and attempts to be smart: 'the sum of the square roots of any two sides of an isosceles triangle is equal to the square root of the other side'.

Here's something to try that will require three of the skills you've seen so far in this chapter:

This is an equilateral triangle with sides of length 2. From each vertex of the triangle a circular arc is drawn, with radius of 1. What proportion of the triangle is shaded?

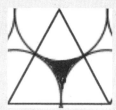

Where to start with this?

Step 1: Area of the whole triangle. The triangle has a base of length 2, but we don't yet know its height. However, if we construct a right-angled triangle by splitting the equilateral triangle in half down the middle, we can find the height we need:

Apply Pythagoras' theorem:
$1^2 + h^2 = 2^2$ (where h is the height of the triangle)
$1 + h^2 = 4$
  $h^2 = 3$
  $h^2 = \sqrt{3}$, or approximately 1.73

So the whole original triangle has width 2 and height $\sqrt{3}$, giving an area of $\frac{1}{2} \times \sqrt{3} \times 2$, which equals $\sqrt{3}$ square units, or approximately 1.73 square units if you prefer.

Step 2: Area of the three circle sectors. Each of these circle sectors is a sixth of a full circle. We know this because each corner (vertex) of an equilateral triangle has size 60 degrees, i.e. a sixth of a full turn (360°). So the three-sixths of a circle essentially make a semicircle.

Area of a semicircle $= \frac{1}{2}\pi r^2 = \frac{1}{2} \times \pi \times 1^2$ (because the circles have radius 1, half the width of the triangle)

So the area of the three circle sectors is $\frac{1}{2}\pi$ (or $\pi/2$), approximately 1.57 square units.

Step 3: Finding the shaded proportion. Finally, to find the proportion of the triangle that is white, divide the area of the semicircle by the area of the triangle:

$$\frac{\pi/2}{\sqrt{3}} \approx \frac{1.57}{1.73} \approx 0.907$$

So the white part is about 90.7% of the triangle, and the black part is about 9.3%. Remember that now.

Looking for *similar* triangles can be very useful. Similar triangles are two or more triangles where one is an enlargement of the other, so both triangles will have the same angles and will have sides in the same ratio. The best way to show this is with a question:

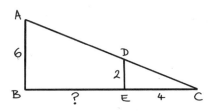

What length could replace the question mark in the above diagram?

The key to cracking this is to view the shape as two triangles, one within the other, the larger triangle having vertices ABC, and the smaller one DEC.

Crucially, we can say that the two triangles are *similar*. In mathematics the word similar does not just mean they look 'roughly the same' – like the actor Alec Baldwin and 13th US president Millard Fillmore – it means that one is literally an enlargement of the other. How do we know triangle ABC is an enlargement of DEC? Well, they both have a right angle in the bottom left corner, and they also share the same angle at C. A triangle only has three vertices, so if two angles are the same, all angles are the same.

Once we've realized the two triangles are similar, we're in business. The height of the larger triangle, AB, is three times that of the smaller triangle, DE, so the width of the larger triangle (BC) must also be three times that of the smaller triangle (EC, which we know is 4). That means BC is 12, so the question mark can be replaced by **8**.

If you couldn't do all that in your head, you might perhaps have written something like this:

$$\frac{BC}{6} = \frac{4}{2}$$

$BC = 12$

Missing side (BE) = 12 – 4 = 8.

It's also worth considering how large triangle ABC is compared to DEC. The area of a triangle is half that of a rectangle, i.e. $\frac{1}{2} \times$ base $\times$ height, so the larger triangle has area 36 square units, the smaller has area 4 square units. Notice that the lengths of the larger triangle are three times that of the smaller triangle, but the area is *nine times larger* (because the height is 3 times more, the width is 3 times more and $3 \times 3 = 9$). Here's a lovely problem to practise your similar shape skills on:

#48

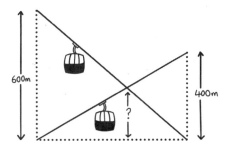

Two cable cars are linked from ground level to the top of two mountains, one 400 m high and one 600 m high, as shown in the diagram. At what height do the cable cars cross?

I find this problem especially enticing because it really feels like it should matter how far apart the mountains are, yet it doesn't! Here are a few scale diagrams to attempt to convince you of that.

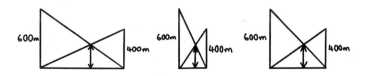

First I will add some labels to the original diagram for clarity:

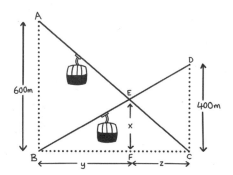

Now the key to cracking this is to realize that triangle ABC is similar to triangle CEF, and that triangle BCD is similar to BEF. If they are similar, that means that the ratio of the height to the width is the same in each triangle:

$$\frac{BC}{600} = \frac{CF}{EF} \quad \text{and} \quad \frac{BC}{400} = \frac{BF}{EF}$$

Substituting the lengths x, y and z from the diagram for the lengths EF, BF and CF, we can write this as:

$$\frac{y+z}{600} = \frac{z}{x} \qquad \frac{y+z}{400} = \frac{y}{x}$$

These two results can be rewritten (by some relatively simple manipulation) as:

$$x(y+z) = 600z \quad \text{and} \quad x(y+z) = 400y$$

Notice that the left-hand side is the same in both cases, which must also make the right-hand sides the same:

$$600z = 400y$$

which simplifies to:

$$3z = 2y$$

In other words, however long the base is, it's split into a 2:3 ratio, where y is slightly longer than z. There are various ways to finish off the job, but my favourite is go back to the last line that contained x, y and z:

$$x(y + z) = 400y$$

We need to boil this down to the point that it only contains x, which currently seems very far off! But if we multiply both sides by 3, something very neat happens:

$x(3y + 3z) = 1200y$ (and remember that $3z = 2y$)
$x(3y + 2y) = 1200y$
     $x(5y) = 1200y$ (finally, divide both sides by 5y)
       $x = 240$

Regardless of how far apart the two mountains are – whether 10 metres or a kilometre – the cable cars will always cross at a height of 240m. Looking for similar triangles can help you to solve a problem that seems to have insufficient information given. Hopefully we're now ready to once again consider the question from the start of the chapter:

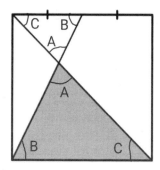

The shaded triangle is similar to the triangle directly above it, with the angles corresponding in the way that I have labelled. Since the shapes are similar, and the shaded triangle is twice as wide as the unshaded one, it follows that the height of the shaded triangle must also be twice that of the unshaded one. This means that the shaded triangle reaches $\frac{2}{3}$ of the way to the top of the square. Now, if we take the base of the shaded triangle to be 1, then its height is $\frac{2}{3}$ and its area is $\frac{1}{2} \times$ base $\times$ height $= \frac{1}{2} \times 1 \times \frac{2}{3} = \frac{1}{3}$. The square has area 1 (because its dimensions are 1 × 1) so the shaded triangle takes up a third of the square.

The 'Pink Triangle' (alas, this book is printed in black and white – please feel free to imagine the shaded triangle in pink) was created by Ed Southall, whom you might remember from the borderline-trolling gorilla factorial question from Chapter 4. Ed's pink triangle is one of his most well-known creations, gaining press coverage on various UK news websites before being picked up as a story all around the world, including New Zealand, Russia and the United States. Oddly though, as Ed explains, before being picked up as a newsworthy story it actually wasn't all that popular a puzzle.

'I think it was *The Sun* that published it first, then the *Daily Mail* featured it, and then it went crazy. But although all the articles described it as a 'viral' puzzle, the original post on Twitter only got about 200 likes. It's almost as if it became a viral question simply because the news outlets decided it was a viral question.'

The legacy of this slow-burning classic dwarfs its initial public reception, perhaps making the Pink Triangle the *Shawshank Redemption* of geometry puzzles.* Ed's teasers often spark a lot of interest on social media, with puzzlers scrambling to submit their solution before others do. Being a geometry Luddite, I'm fascinated to know where he gets his crazy ideas from.

'Sometimes I start a problem with the method of solution in mind, but often I'm just messing about with a computer geometry package; playing around with regular shapes, intersections, midpoints etc. I constructed the 'pink triangle' picture from messing around with a square and this interesting triangle and thinking the area looked obvious, but then I thought – is it obvious? I could see maybe two or three ways that it could be done, so I posted it. That's the big appeal for me – not in trying to outfox people with a problem but seeing the various ingenious methods that people come up with.'

Ed has compiled the ten smartest solutions – yes, there are many more than ten ways of doing it! – having taken the time to work through dozens and dozens of methods that have been posted on Twitter or emailed to him.† He even broke the 'first rule of the internet' and read the comments on news websites, always hazardous terrain.

'I think what's so appealing about the question is that there's so little going on, and it does look like about a third is shaded. It's clearly

---

* And, like Andy Dufresne, the novice puzzler may find they have to tunnel through a lot of shit to get to the eventual redemption.

† https://pure.hud.ac.uk/ws/portalfiles/portal/18775330/Approaches_to_the_Pink_ Triangle_Problem.pdf

between a quarter and a half, so… probably a third. And for some people, that's enough! What's fascinating to me is that lots of people don't actually gain any more satisfaction from finding out *why* it's a third. They'll reply to an elegant proof by saying, "Yeah, I said it was a third at the start." I think that's the real sign of a mathematician – whether they care *why* it's a third.'

You might recall this same behaviour from the $8 \div 2(2 + 2)$ problem from a few chapters ago, with people queuing up to declare the answer either 16 or 1, with no willingness to actually try to understand where the two possible results come from. To reference iconic 1990s sci-fi classic *The Matrix*, many people are happy to take the blue pill and just accept that the answer is a third. But only the true mathematician is prepared to take the red pill, see *why* the answer is a third and how far down the rabbit hole goes.

I do feel, however, that there's one aspect to the Pink Triangle's viral success that Ed might not have considered: the pinkness. The grey triangle, or the turquoise triangle, just don't seem to have the same appeal. 'Yes,' agrees Ed, in typically self-deprecating style. 'If it was the yellow triangle it might have only got 100 likes.'

When not creating gloriously frustrating yet ultimately enlightening online content, Ed is a lecturer and trainer of early-career maths teachers. I asked if he had a particular geometry problem in mind that generated that beautiful 'ah!' moment that all teachers would hope to encounter in the classroom, and I'm delighted that he was happy to oblige:

#49

Draw a square (freehand is fine, it doesn't need to be perfectly accurate).

Put a dot anywhere in the square, and join it to the four corners of the square (again, freehand is fine).

Your square should now be split into four triangles. Shade two alternate triangles, i.e. two triangles that do not share a common edge. How much of the square is shaded?

Here's one way you could draw it, but you almost certainly drew yours differently. Crucially though, it doesn't matter – we will still get to the same answer.

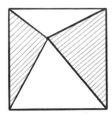

Like so many geometry puzzles – certainly some of the best ones – there's a long and methodical way to get the right answer, but also a quicker, neater way that involves spotting a clever twist. When I was a child one of my favourite storybooks was Terry Jones' *Fairy Tales* – yes, he of *Monty Python* and *Labyrinth* fame (though I didn't know that at the time). In one of the stories, a girl is invited by a fairy to follow her to the goblin city, the fairy teasing her with the following refrain:

Short or long to Goblin City?
The straight way's short
But the long way's pretty…

Solving geometry problems is very often the other way round: the straight way's long but the short way's pretty! The difficulty, of course, is spotting the short way, and this comes with practice.

Long way

Drawing the following horizontal and vertical lines allows us to find the area of all four triangles. The shaded triangles have heights a and b but both have width $c + d$. The unshaded triangles have heights c and d but both have width $a + b$.

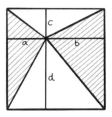

Using our old favourite, area of a triangle $= \frac{1}{2} \times$ base $\times$ height, and some basic algebra, we get:

Shaded area: $\frac{1}{2} a(c + d) + \frac{1}{2} b(c + d) = \frac{1}{2} (a + b)(c + d)$
Unshaded area: $\frac{1}{2} c(a + b) + \frac{1}{2} d(a + b) = \frac{1}{2} (a + b)(c + d)$

It transpires that the shaded and unshaded areas are equal, so both must take up half of the square.

Short way

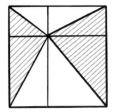

Draw in the same horizontal and vertical lines, but this time there's no need to label any of the lengths. Notice the square could now be considered as four rectangles that meet at the point you drew, and that each of the four rectangles is split in half with a diagonal line. Each rectangle is half-shaded, so the total shaded area must equal the total unshaded area.

There's something so very satisfying in finding a neat shortcut and watching the apparent difficulty of the problem dissolve away, similar once again to the moment a magic eye puzzle changes from fuzzy mess to crystal clear 3D image (probably: I've been staring at them since 1996, still to no avail).

A student of mine suggested an even quicker way to the correct answer, though this approach does include a meta-knowledge of the way these puzzles usually go. I asked you to shade two triangles, but I didn't specify which two to shade. That means that in two parallel universes you could have shaded top and bottom or left and right. But the result I required should be independent of which triangles you shaded, so one might guess with some certainty that the answer is going to be that shaded and unshaded are equal.

Here's another beautiful example of the 'ah!' moment:

#50

The circumference of this circle is split equally into 12 sections.
What proportion of the circle is shaded?

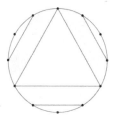

Credit: Twitter/HenkReuling

This problem was posted on Twitter in 2018 by Dutch maths teacher Henk Reuling and went decidedly un-viral, but it has a beautiful 'ah!' twist when you see it.

Long way

The white areas are easier to find than the shaded areas, since the central part is an equilateral triangle and the outer white parts are 'segments' of the circle. It's still rather fiddly to get to the final answer this way, so I've put it at the back of the book, but the answer you arrive at is very satisfying: a half. Which implies that there's probably some neat quick way...

Short way

As noted, and illustrated, by Twitter user Ignacio Larrosa Canestro (@ilarrosac), the shading can be slightly tweaked by 'moving' the three white sections on the outside clockwise by one dot.

Credit: @ilarrosac

The circle is now split into six identical triangles, half of which are shaded; and six small segments around the outside, half of which are shaded. So the circle is half-shaded. Gorgeous!

Henk's 12-point circle was recommended to me by a legend of the 'ah!'-moment geometry problem, Catriona Agg. Catriona is a UK-based maths teacher who creates charming felt-tipped geometry problems that are hugely popular on Twitter. Here's a classic example of what she does so well:

#51
A square and four half-circles. What proportion of the total area is shaded?

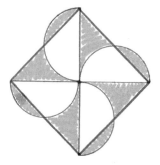

Credit: Catriona Agg

Goodness knows what the long way is to do this; it feels like it would be horrific. But the short way is lovely – each of the shaded sections outside the square can be moved inside the square in the following way:

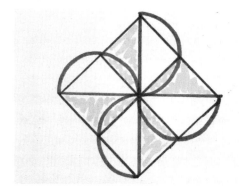

Credit: Catriona Agg

Now we can clearly see that the square is half-shaded. Delicious!

Again, as a relative geometry novice, I have to know: what comes first, the question or the 'ah!' moment?

'Oh, the planning is completely random. I just doodle and hope that a puzzle falls out,' says Catriona (what is it with these self-deprecating geometers?). 'It's just scribbles everywhere and I hope that eventually something comes of it.' Like Ed Southall, she too sometimes feels a disparity between what feels like a 'good' puzzle and what turns out to be a popular puzzle. 'Sometimes I'll be drawing and redrawing the same picture for days, convinced that there's a really neat solution, and it won't be that popular. Then sometimes I'll do something that feels really derivative and it will get thousands of likes! But it all starts off in one of my doodle books.'

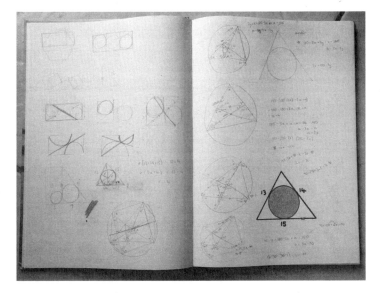

Credit: Catriona Agg

Catriona is such an advocate of the 'ah!' moment that she describes it as the very thing that separates a puzzle from a mere maths question. Not that all of her students agree. At this point she speaks in the voice of a frustrated student: 'All this time we've been working on it and you could have shown us how to do it in one minute!' Sounds like people wanting to take the blue pill to me.

Usually there are various ways to crack a trademark Catriona puzzle, such as the following example:

#52

A rectangle and a square. What is the area of the rectangle?

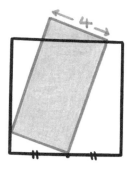

Credit: Catriona Agg

To which the following represent just a tiny sample of the responses:

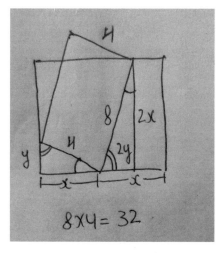

Credit: @Expert_Says

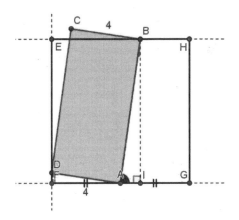

Credit: @dandriana

Credit: @AmareshGS1

Some people have used trigonometry – which you may be less familiar with – but it's not necessary if you spot the similar triangles, as the aptly named '@Expert_Says' has done above, and '@dandriana' has made even clearer. They have constructed a triangle on the right-hand side that is similar to the triangle in the bottom-left, and since one is a 2:1 enlargement of the other, the rectangle must be twice as long as it is wide, giving an area of **32** square units.

Yet of course there is an even quicker way! Since no angle is given for the elevation of the rectangle – i.e. the angle at which the rectangle is leaning – it should be the case that the same diagram could be drawn with a much shallower angle, as shown in the @AmareshGS1 image above.* If the angle could be made shallower without changing the question, following this argument to its natural conclusion means the angle could be made to be zero. If that were the case, the rectangle would sit exactly within half of the square, again giving an area of 32.

I find the veritable smorgasbord of methods in the replies to Catriona's tweets to be one of the beacons of positivity in the hostile environment that social media can so often be. Come to one of her puzzles 48 hours after it's been posted and you'll find dozens of approaches, gifs and animations, novices asking for help (and many people kindly obliging), and – make sure you're sitting down – polite and reasoned debate. On the internet! Whatever next?

There does, however, seem to be one downside of being a teacher as well as a viral puzzle setter. 'I had one class a few years ago who got very good at looking things up on Twitter,' Catriona laughs. 'They'd be secretly looking at their phones under the table or going on my Twitter feed in advance to try and get the answers. I had to start giving them

---

* His response was actually a moving image, but I can't even show a pink triangle here, let alone a video. Find it and all these responses at kyledevans.com/mathstricks

problems *before* I'd set them on Twitter, which they thought was very unfair!' I hope these students are still looking at Catriona's page since leaving her class; I highly recommend it as an ocean of geometric calm in a world of uncertainty.

#53

Arrange four points so that there is a distance of exactly 2 metres between any pair of points.

This is a lovely tool for getting people to think outside the box. It's quite intuitive to solve the same problem for three points: an equilateral triangle, i.e. a triangle where each point is 2m from both other points, so that all three sides of the triangle are 2m long. But where can the fourth point go? Anywhere you place it will necessarily be closer to some points than others, surely?

The question is impossible, but only if you moor yourself to a two-dimensional universe. If you allow yourself the luxury of three dimensions – go on, spoil yourself! – we can place a fourth point off the plane that contains the first three points, forming the four vertices of a tetrahedron:

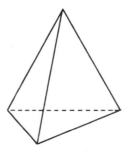

If every edge on this tetrahedron is 2m long, then every vertex, or corner, is exactly 2m from every other. This puzzle suddenly became more real-world relevant as a result of some poorly drawn social distancing guidance that was spotted during 2020:

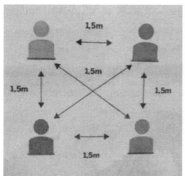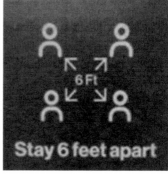

(Source unknown)

As we've seen, it's impossible for four people to stand in such a way, so that everyone is the same distance from everyone else, unless they were to construct some kind of socially distanced tetrahedron structure. Which, to be fair, would probably have been just as useful a way to spend a lockdown day as baking more banana bread.

This type of social distancing-based geometry problem might seem silly and trivial, but it did have real-world implications when I had to work out how best to pack students into classrooms when they returned to college after lockdown in late 2020, while maintaining social distancing due to the still ongoing coronavirus pandemic. Creating a two-level structure with a tetrahedral lattice network between seats in the classroom was apparently beyond our budget, so we were restricted to working with tables and chairs on the floor. Still, it was nice to be able to use some actual real-world maths in the workplace.

'Am I right in thinking rows and columns would be the best way?' was the question I was first asked by the college caretaker. Well, the trick here is to understand that 2-metre social distancing is essentially equivalent to everyone in the class sitting in the centre of a circle with a 1-metre radius, and packing as many of these circles into a rectangular room as possible. Let's go with a rectangular array of rows and columns, as the caretaker first suggested:

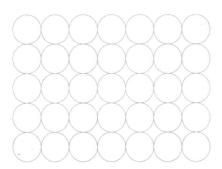

If you can imagine the centre of each circle being where a person is sitting, this would look like classic 'exam' style seating, with horizontal and vertical distancing of 2 metres between desks. The trick with *circle packing*, as this problem is known, is to fit as many circles into the plane as possible. To do this, we need to minimize the area between the circles: a smaller area between the circles means more of the plane is made up of circles, which in our social distancing analogy means more people packed into the room. First let's consider how efficiently the room is being filled in the above example:

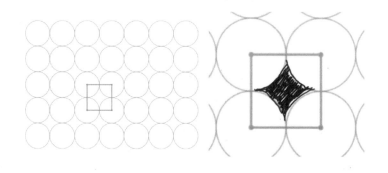

We could imagine that the plane is tiled in squares as shown above, so that if we can find the proportion of the square that's made up of circle parts (the area that isn't shaded), we can similarly find the proportion of the whole room that's being utilized.

Since the circles have a radius of 1m, and each edge of the square is made up of two of these, the square is 2m by 2m, giving an area of 4m². There are four quarter-circles inside the square, which is equal to one circle. Each circle has area πr^2, which in this case is just π, because the circles have radius 1. Therefore the proportion of the square that's white is $\pi/4$, which is about 0.785. The room is 78.5% covered in circles.

Could we do any better? It turns out we can, as the caretaker was absolutely delighted to discover.

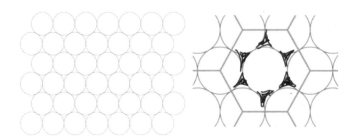

Arranging the circles in a so-called *hexagonal* packing allows for a smaller shaded area. In fact, split the hexagon into six identical equilateral triangles and you'll find that you've already answered this question back in the box on page 154.

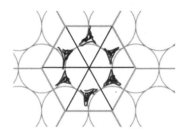

The hexagon is made up of six identical triangles, and we've shown each triangle to be about 91% covered, so the entire packing design is about 91% covered. This is the optimal packing for circles in a plane, as first investigated by Gauss and Lagrange around the turn of the nineteenth century, though it took some 150 years for the hexagonal packing to be proven as the best.

'I feel like this conversation's been going on for 150 years,' said the tired caretaker. 'Can we just go for rows and columns?' So we did.

CONCLUSION

A better viral maths future?

Having spent the best part of a year thinking a lot more about viral maths puzzles than I would ever have expected to, I find myself torn. It has struck me that so many of the puzzles and viral tricks I've featured took off around 2015, just a year before the Western world would be thrown into one of the most polarized periods in its history. Perhaps puzzles like Cheryl's birthday, 8 ÷ 2(2 + 2), #thedress and brainstorm / green needle were sent to prepare us for the bitter, partisan new world of social media we were about to live through? On one hand, I feel compelled to try to create a more socially responsible, ethical form of viral puzzle that doesn't resort to pitting people against each other for cheap shares and comments.

But then there's the devil on my other shoulder that craves those likes and retweets, and knows, deep down, that people just won't be interested in a viral puzzle with one agreed neat solution. In an attempt to resolve this restlessness, I decided to launch a scientific study. Over a Bank Holiday weekend, I launched two emoji equation-style puzzles, one at noon on Saturday and one at noon on Sunday. The first was my attempt at an ethical, responsible, good-hearted problem; for the second I gathered everything I've learnt in the book and put it to the worst possible use, aiming purely for argument and division (both types).

#54: Angel puzzle

💜 + 💜 + 💜 = 24
💜 × 😊 + 💜 = 32
💜 − 👍 + 😊 = 9
💜 ÷ 👍 (😊 − 👍) = ?

Because this is a kinder, more morally pure form of viral puzzle, I also decided to adorn it with the slightly twisted clickbait heading: 95% OF PEOPLE CAN SOLVE THIS!

The solution: From the top line, 💜 is clearly worth 8. In the next line I have used multiplication and addition in the order that even people working from left to right should get the right answer, since multiplication is to the left of addition anyway:

💜 × 😊 + 💜 = 32
8 × 😊 + 8 = 32
8 × 😊 = 24 (subtracting 8 from both sides)
😊 = 3 (dividing by 8)

The third line involves only addition and subtraction, so we should be fine here too:

💜 − 👍 + 😊 = 9
8 − 👍 + 3 = 9
11 − 👍 = 9
👍 = 2

Now we simply have to plug this all into the bottom line:

♥ ÷ 👍 (😊 – 👍)= ?

$8 ÷ 2(3 – 2) = ?$

Argh! Wasn't this the exact type of mischievous trick question that caused so much misery in the middle of the book? Well, almost, but with a twist of kindness. Observe...

If we multiply first:	If we divide first:
$8 ÷ 2(3 – 2)$	$8 ÷ 2(3 – 2)$
$= 8 ÷ 2 × 1$	$= 8 ÷ 2 × 1$
$= 8 ÷ 2$	$= 4 × 1$
$= 4$	$= 4$

Whichever way you read it, the answer comes out the same! This only works when we force the part in the brackets to equal 1 (or –1 if you're feeling adventurous). Isn't that nice? A seemingly divisive emoji equation problem, but with the twist that everyone can agree on its answer. Now, I unleash my dark side:

#55: Devil puzzle

🧍 + 🐕 + 🐟 = 6

🐜 + 🦞 = 14

(🐝 – 🐌 × 🐕) ÷ 🦎 = ?

The solution: Surely there isn't enough to go on here?! Each emoji appears only once. I have been trained in the dark arts of clocks and banana emojis though, and I know that the public are now primed to scratch a little deeper below the surface. What do a man, dog and fish have that adds up to 6? Or an ant and octopus that adds up to 14? It's legs, isn't it? So the last line translates to:

$$(\text{🕷} - \text{🐛} \times \text{🐎}) \div \text{🐛} = ?$$
$$(8 - 2 \times 4) \div 0 = ?$$

Which takes us to $0 \div 0$. What's that then? Zero divided by zero? Here are two lines of thought:

- It's 0, because there is nothing to 'share out'. Division, in practical terms, means sharing an amount, or 'dividend', into equal shares. If the dividend is zero, that means there is nothing to share out. If Hannah has no sweets, it doesn't matter how many people she needs to share them between: nobody gets any sweets!
- It's 1, because the dividend and the 'divisor' (the amount you're sharing between) are equal. $5 \div 5 = 1$, $42 \div 42 = 1$, so $0 \div 0 = 1$.

You probably won't be too surprised when I reveal that neither of these answers is correct: actually, zero divided by zero is *undefined*; maddeningly, its limit can behave differently depending on the context you find it in. There is literally no answer to this puzzle. I concede this feels like a bit of a cop-out, but, honestly, it's best to avoid ever discussing zero divided by zero if you can help it. A bit like England's performance at Euro 2016: just don't mention it, and hopefully it will go away.

How about my scientific trial: which do you think was more popular – angel or devil puzzle? This is really a good news / bad news situation, the good news being that they were almost identically popular, each being viewed by about 8000 Twitter users over the course of the weekend. The bad news is that neither of these tweets came anywhere near viral territory: for reference, my cheese pun from the introduction – still far from a viral sensation – was viewed by 314,000 people (yes, pi-hundred thousand).

I should have known better than to attempt any kind of viral formula though, because if I truly learnt anything about virality in recent years, it's that the only real rule is that there are no rules. This book was written entirely during the Covid-19 pandemic of 2020/21, a time during which vast swathes of us began working from home. Our lives became smaller overnight, with our main point of contact with our colleagues, peers, families and the outside world becoming – more than ever – social media. This situation led to two particularly weird and wonderful viral maths sensations that give me hope for a better viral maths future.

Social media at its worst can often be reminiscent of the playground, with all the name-calling, bullying, tale-telling and posturing that we all should have left behind when we were 12. A particularly nasty example of this began to bubble up when American teenager Gracie Cunningham posted a short video on TikTok, questioning whether math(s) is real while applying her make-up.

I was just doing my make-up for work and I just wanted to tell you guys about how I don't think math is real. And I know that, like, it's real because we all learned it in school or whatever. But who came up with this concept? And you'll be like 'Pythagoras.' But how? How did he come up with this? He was living in, like, the, I don't know whenever he was living, but it was not now, where you can, like, have technology and stuff, you know. Like, he didn't even have plumbing. And he was, like, 'let me worry about $y = mx + b$.' Which, first of all, how would you even figure that out? How would you, like, start on the concept of algebra? Like, what do you need it for? You know, cos, like, I get an addition.

Like hey, if I take two apples and add three, it's five, you know? But how would you come up with the concept of, like, algebra? Cos, what would you need it for? Know what I mean? What would you need it for back then? You didn't need it, so why come up with it?

This video gradually racked up a million views over a couple of weeks, and then Twitter got involved. It was reposted on Twitter – not by Gracie, but by another user – with the caption 'this is the dumbest thing ive ever seen', where it collected more than 13 million views and hundreds of thousands of likes in under a week. It seemed that people took great joy in revelling in the stunted musings of an American teen who can't get through a sentence without a 'like' or 'stuff'. Elitism and belittlement followed in droves; I won't repeat any of what was written here. But having a worldwide 'pile-on' on a curious 16-year-old girl suddenly seemed to be fair game.

And then… something changed. Some very qualified mathematicians and physicists weighed in, backing up Gracie's scepticism. Actually, why was algebra invented? What was the first event that tipped a mathematician from not needing algebra to needing algebra? Some academics read out Gracie's video, with a little of the teen-speak removed, to show that when presented by a person of supposed stature, the questions are not stupid or naive at all.

This kind of question is integral to the mathematical experience. In fact, the mathematician who wants to know *why* maths works, who wants to know *why* the pink triangle covers a third of the square, or *why* the 1089 trick works, is exactly the kind of person who would question if maths is really real at all.

Whether maths was discovered or invented has long been a contentious subject that mathematicians love to discuss when they're avoiding doing proper work, which is often. Einstein remarked (in a TikTok video in 1936), 'How can it be that mathematics, being after all a product of human thought which is independent of experience, is so admirably appropriate to the objects of reality?' The mathematics that humans have come to use fits the workings of the universe so well, surely it is somehow innate and not a human construction? Conversely, many would argue that the only reason our mathematics fits the physical world so well is that we invented it to do so, and that there are many aspects of the physical world that mathematics cannot explain, despite thousands of years of trying. If mathematics is so innate, what's the deal with that?

Or, to put it in short, *is math real*? I first became aware of Gracie's video during the positive fightback against the initial backlash, which was much greater than the original negativity and lasted much longer. This is probably one of the only times that I can remember mathematical philosophy making it into the public consciousness. People around the world were engaged in very serious arguments about whether or not maths is 'real', and it all started with a 16-year-old girl applying her make-up.

Gracie was happy to share her side of the story with me, once she was certain that I was in her corner and she wasn't about to receive another online dragging. 'It was definitely weird to see so many people talking about me in that way, especially when I didn't even put the video on Twitter – someone else did. There was a lot of hate which took a while to turn round into support.'

Has it had any long-lasting effect? 'OMG yes. I've mostly been able to move on from it, but it's still a huge online joke among my friends.'

While Gracie and #mathisntreal eventually ran out of momentum, on this side of the Atlantic our own understated viral maths superstar was making waves of his own, somewhere in a loft in Surrey.

One day, when absent-mindedly scrolling through social media on another bland and ordinary lockdown day, I came across a link a friend had posted. 'I don't know what this says about what lockdown has done to me', my friend wrote, 'but this 25-minute video of a man solving a sudoku is the most thrilling thing I've seen all year.' Needless to say, as a logic puzzle enthusiast, I was intrigued.

The video in question was from the YouTube channel *Cracking The Cryptic*, in which two experienced and exceptionally talented puzzle enthusiasts, Simon Anthony and Mark Goodliffe, solve immensely difficult logic puzzles live. The puzzle in question was the 'miracle sudoku', in which Anthony solves a sudoku given only two starting digits and a whole load of other rules. Yes, *two* starting digits. (It was an anti-knight, anti-king, non-consecutive sudoku,* and if you're interested in what the hell that means, or even trying one, there's an example at the back of the book – which is quite close now.)

At first Anthony is absolutely certain that he's being trolled by the puzzle setter, Mitchell Lee, before slowly realizing that, actually, it might just be solvable. Here are some highlights from the video; imagine them being delivered in an increasingly agitated and very, very British fashion by a man who realizes, minute by minute, that something incredible is unfolding.

3:45   Simon uploads the puzzle.
3:50   'Right. He's got to be joking. There's no way that… well it might have a unique solution but it's not going to be findable by a human being.'
4:18   '… although I can place a 1 in this central box…'
6:18   (Simon has placed a few 1s and 2s.) 'OK, we may have to actually take this seriously.'

---

* In fact this set of constraints is now simply called a 'miracle' sudoku: the puzzle in the video created the genre.

9:52   'You are kidding…'

10:55 'Oh my goodness. What is going on with this puzzle?'

12:12 (Most of the 1s and 2s placed.) '… this is probably going to be impossible now…'

12:50 '… this is just not going to solve…'

17:30 '… this is just staggering, this is *absolutely* staggering…'

18:20 (Simon starts placing some 3s.) 'This is magic. We are watching magic unfold here.'

19:23 (Simon has placed all the 1s, 2s and 3s.) 'This is going to solve, isn't it? I am pretty dumbfounded, I have to say.'

19:50 '… I'm not sure I have the adjectives to describe what's going on here. This is like the universe is singing to us…'

21:40 (4s and 5s complete.) 'This is something extremely special.'

23:50 (Simon starts placing 8s and 9s.) 'We are watching something extremely special here. Mitchell Lee has come up with a work of sublime genius.'

The video went huge, exploding outside of the world of YouTube puzzle aficionados. From vloggers and influencers to novelists and Hollywood writer–directors,* no one could get enough of this mild-mannered man going quietly insane over the best puzzle he'd ever seen. I met Simon six months after the dust had settled to ask what it felt like to realize the 'miracle' sudoku had truly gone viral outside of the world of online puzzle-solvers.

'It's like a drug,' says Simon, clearly tapping right back into the excitement all these months later. 'When I started the channel I was still working in the city and I'd get an email every time the channel

---

* Bestselling novelist Audrey Niffenegger and Peter Howitt, the writer–director of *Sliding Doors*, have both publicly stated their intention to build characters based on Anthony's loft-based exploits.

got a new subscriber. I'd think it was a good day if I got an email – one single new subscriber! At the height of the channel's popularity this summer we were getting thousands of new subscribers every hour. We were getting new subscribers and comments from all around the world, and featuring in news articles around the world. It's an extraordinary and rather surreal feeling.'

All this acclaim for a YouTube channel that is quite literally a man solving a sudoku from a loft in Surrey. To what does Simon credit the appeal of the channel? 'We broke all the rules. Everyone says not to go beyond ten minutes, and to keep it snappy. Anyone who's seen the channel knows that we ramble on a bit, and we certainly aren't snappy talkers. Once I get into my flow I lose all track of time and become consumed by the puzzle. Maybe the viewers can do the same, and it gives them a way to block out the world for half an hour.'

To me, watching *Cracking The Cryptic* is like watching a master sportsperson at the top of their game. Elite puzzlers who are watching might spot the occasional slip – Simon says he routinely beats himself up about the moves he *should've* made – but essentially it's like watching Lionel Messi play football, or Serena Williams play tennis. When at university myself, I always loved watching lectures from immensely intelligent mathematicians; there's something so thrilling about watching someone who is an expert in their field do immensely difficult mathematics, and holding on for dear life as you try to keep up. In the case of some of my lecturers it didn't even matter that they themselves were terrible communicators, as long as they were passionate about what they were doing. Though being a poor communicator is not a slight that could ever be levelled at Simon, who is able to explain his methods with welcoming clarity, to an audience ranging from fellow world-beating puzzlers all the way down to complete novices.

I could easily see him as a fellow teacher, a route it transpires he considered before leaving the city to dedicate himself to *Cracking The Cryptic* full-time. Is he pleased he took the plunge and grasped his own *Sliding Doors* moment? 'Oh absolutely. I was doing a job that I despised, and now we've built this online community that's just staggering. I'm so glad we did it.' Looking at their most popular videos – the 'miracle' has 2.5 million views, their most popular video has over 5.5 million – it seems the puzzle-solving public is glad they did too.

At times in this book I've been down on social media, but what joy to live in a world where a man solving a sudoku online for half an hour can be a viral hit. What a world to live in where bullies and trolls queue up to dunk on a 16-year-old girl, only for world-leading experts in mathematics with huge online followings to come to her defence. I know the geeks are meant to be inheriting the earth; how life-affirming to see that now might finally be the time.

Recently I discovered the following quote by the writer and sustainability campaigner Anna Lappé:

> *Every time you spend money, you're casting a vote for the kind of world you want.*

I can't help but draw a parallel to our online lives, and all the comments, shares and 'likes' we make on social media. There's so much good, wholesome, mind-expanding maths content on social media; no one needs to waste their life arguing about whether an emoji equation has three or four bananas or if the clock hands are on 2 o'clock or 3 o'clock. It's great that social media has helped discover a wellspring of people who, despite their protestations, take an interest in maths and want to do more of it. Now we just need to keep pointing them towards the good stuff.

SOLUTIONS

Introduction

The cheese tweet – page xiv

Most people are used to measuring angles in degrees: 180° for half a turn, 360° for a full turn, etc. But degrees are a construct, used mainly because 360 is a very divisible number: you can divide it cleanly, with no remainder, by 2, 3, 4, 5, 6, 8, 9, 10, 12, etc.

A more mathematically useful, though perhaps less 'pleasing' way to measure angles is in radians, where one radian is defined as the angle created when a circle sector has the same radius as arc length:

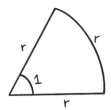

It's also true that exactly π radians make half a turn, where π is a number just slightly greater than 3 that starts 3.14159… and then goes on endlessly. The piece of cheese costs £π, and happens to be pretty much a perfect quarter-turn. By this reasoning, half a whole Brie (π radians) would cost £2π, so the cheese appears to cost £2 per radian.

(I did promise the joke would be even less funny when I explained it.)

1 THIS ONE COOL MATHS TRICK WILL BLOW YOUR MIND: Maths tricks and 'life hacks'

Try it yourself – page 4

(a) 73% of 10 = 10% of 73 = 7.3

(b) 12% of 25 = 25% of 12 = 3

(c) 16% of 75 = 75% of 16 = 12

(d) 44% of 5 = 5% of 44 = half of 10% of 44 = half of 4.4 = 2.2

(e) 13% of 25 = 25% of 13 = half of 50% of 13 = half of 6.5 = 3.25

The ten-numbers-multiplied-together trick – page 17

When choosing ten numbers, each of which can take the values 1–9 (i.e. 9 possibilities), there are 9 choices for the first digit, then 9 choices for the second digit, then carry this on for all ten digits, giving 9^{10} potential outcomes, or about 3.5 billion (though we don't really need to know that).

It's probably easiest next to count all the situations when the trick *doesn't* work. One way this could happen is if every digit selected is a 1, 2, 4, 5, 7 or 8, i.e. six possibilities. Following the same logic, there are 6^{10} ways this could happen. We also need to count all the cases when just one of these values is a 3 or 6, but all the others aren't a 3, 6 or 9.

To do this, consider the first value chosen is a 3 or 6, so two possibilities. After that, every other of the nine values chosen must be a 1, 2, 4, 5, 7 or 8, so 6^9 possibilities. So far we have 2×6^9. However, this only counts the situation when the first chosen digit is 3 or 6. The 3 or 6 could have happened in any of the ten possible positions, so we actually need $10 \times 2 \times 6^9$.

So the total possible 'fail' situations is $6^{10} + 20 \times 6^9$, which is about 262 million. There are 262 million ways this trick could go wrong! Yikes! But as a proportion of 3.5 billion that is about 7.5%, in other words the trick will almost always succeed.

In reality, my anecdotal experience suggests the trick actually works more often than 92.5% of the time, probably because humans are atrocious at choosing ten random numbers, and instead tend to spread their numbers evenly across 1–9 so that most numbers get picked, i.e. a 9 or a couple of 3s or 6s are picked.

2 IT WAS DIFFERENT IN MY DAY: Pre-internet viral maths

Four more from the playground – page 34

1. At 2 o'clock you can't start timing until the first chime, and you stop timing on the second chime, so there are 2 seconds between chimes. For 3 o'clock there will be 2 seconds between the first and second chime, and 2 seconds between the second and third chime, so the answer is **4 seconds**.

2. The only place your house can be is on the North Pole, so the bear is **white**. As if there weren't already enough reasons for

humans to stop overheating the planet, this riddle will no longer work if we permanently destroy the polar bears' habitat.

3. The bus driver's name is **whatever your name is**, since the first line of the riddle says *you are a bus driver*.

4. **Carrot!** I can't seem to find any decent evidence on whether people actually say carrot, and if so whether carrot is just the most common response regardless of the maths beforehand. It always seemed to work back in the day though. I'd be very happy for anyone reading this to carry out a trial and let me know.

3 BACK TO SCHOOL: Viral exam questions and classroom conundrums

Heather's beads – page 53

There are n beads in total, and to start with 6 of these are black, so $n - 6$ are white. After the first white bead is taken there will be $n - 1$ beads remaining in total, and $n - 7$ of these will be white (one fewer than in the previous step).

The probability that Heather picks two white beads is

$$\frac{n-6}{n} \times \frac{n-7}{n-1} = \frac{1}{2}$$
$$\frac{(n-6)(n-7)}{n(n-1)} = \frac{1}{2}$$

$$2(n-6)(n-7) = n(n-1)$$
$$2(n^2 - 13n + 42) = n^2 - n$$
$$n^2 - 25n + 84 = 0$$

Try it yourself: Neha's sweets – page 54

$$\frac{3}{n} \times \frac{2}{n-1} = \frac{1}{7}$$

$$\frac{6}{n(n-1)} = \frac{1}{7}$$

$$42 = n(n-1)$$
$$42 = n^2 - n$$
$$0 = n^2 - n - 42$$

So $a = -1$ and $b = -42$. The only positive whole number that satisfies this equation is 7, so there were 7 sweets in the bag originally.

The crocodile and the zebra – page 56

You may prefer to look at the solution for the 'optimizing the sheep pen against the wall' problem (page 58) before tackling this one.

Time will be minimized when the first derivative of x against time is zero:

$$T'(x) = \frac{5}{2} 2x(36 + x^2)^{-0.5} - 4 = 0$$

$5x(36 + x^2)^{-0.5} = 4$ (multiply both sides by $\sqrt{36 + x^2}$)

$5x = 4\sqrt{36 + x^2}$ (square both sides)

$25x^2 = 16(36 + x^2)$ (expand and rearrange)

$25x^2 = 576 + 16x^2$

$9x^2 = 576$

$x^2 = 64$

$x = 8$

The crocodile should swim to the point where $x = 8$; this gives the optimal minimum time.

Optimizing the sheep pen against the wall – page 58

Let the shorter side of the sheep pen be represented by x. Then the longer side is $36 - 2x$, and the area of the sheep pen is $x(36 - 2x)$. The maximum area will occur when the rate of change of area against x (the derivative of the function $A(x)$) is zero.

$A = x(36 - 2x) = 36x - 2x^2$ (find the derivative of $A(x)$)

$\dfrac{dA}{dx} = 36 - 4x$ (set the derivative to 0)

$36 - 4x = 0$

$4x = 36$

$x = 9$

Note: the function in the top line, which we are optimizing, is a quadratic function. By nature these functions are symmetrical, which explains why we get symmetry when we tabulate the possible outcomes.

Prison problems – page 80

1. The best way is to think of each prisoner, and which guards would turn their lock:

Prisoner 1:   Guard 1

Prisoner 2:   Guards 1, 2

Prisoner 3:   Guards 1, 3

Prisoner 4:	Guards 1, 2, 4
Prisoner 5:	Guards 1, 5
Prisoner 6:	Guards 1, 2, 3, 6
Prisoner 7:	Guards 1, 7

...

Here prisoners 1 and 4 would be able to escape come morning, since their locks have been turned an odd number of times (and hence have changed state from locked to unlocked). All other locks have been turned an even number of times and therefore remain locked. The 'guard numbers' in the right-hand column are actually the *factors* of the cell number in the left-hand column, so the lucky prisoners are those in a cell with an odd number of factors. Which cells have an odd number of factors? The cells that are *square* numbers, i.e. 1, 4, 9, 16, 25, 36, 49, 64, 81, 100.

2. The prisoners should designate one 'counter' who will eventually report back to the warden. If a prisoner who is not the counter enters the room and sees the switch in the 'down' position, they move it to the 'up' position, but they only ever do this once. If, on a later night, they see the switch in the 'down' position but know they've already flicked the light to the 'up' position in the past, they do nothing. If anyone but the non-counter ever sees the light in the 'up' position, they do nothing. If the counter ever sees the light in the 'up' position, they flick it down and add one to their tally. When his tally hits nine, the counter can tell the warden that all prisoners have visited the room. This is not a very efficient or quick process! But it is guaranteed to eventually work.

3. The following strategy will result in nine prisoners being released, and the tenth prisoner having a 50/50 chance of survival. The first prisoner, at the back of the queue, counts the blue hats in front of them and shouts 'blue' if they see an even number and 'red' if they see an odd number. Now the next prisoner can similarly count the blue hats in front of them, and if they count the same number as the first prisoner they'll know they are wearing red. If they count differently to the person behind, they'll know they are wearing the blue hat. If all prisoners listen to everyone behind them, they can similarly ascertain the colour of their own hat. The prisoner at the back of the queue is effectively the martyr for the whole group, but there's no way they could ever have better than a 50/50 chance anyway. Note: this method could be adapted for a queue of any length.

Algebra homework – page 82

When we multiply a value by itself we *square* it, so that $a \times a = a^2$. If we multiply by a again, we get $a \times a \times a = a^3$, and so on. If we summarize in a table:

Calculation	Index notation
$a \times a \times a \times a$	a^4
$a \times a \times a$	a^3
$a \times a$	a^2
a	a^1

Following this pattern, what should go in the next boxes? Notice that every time we move from one row to the next, we are actually dividing by a (because we 'remove' a multiplication by a). So the next left-hand box will simply be 1 (a divided by a) and the right-hand box will be a^0, because the power in the right-hand column is dropping by 1 every time:

Calculation	Index notation
$a \times a \times a \times a$	a^4
$a \times a \times a$	a^3
$a \times a$	a^2
a	a^1
1	a^0

This tells us that anything to the power of zero is 1 (well, other than zero to the power of zero, which is a whole other can of worms, or *undefined* as we call it in mathematics). But what if we go even further? The established pattern is to keep dividing by a on the left, and reducing the power by 1 on the right.

Calculation	Index notation
$a \times a \times a \times a$	a^4
$a \times a \times a$	a^3
$a \times a$	a^2
a	a^1
1	a^0
$\dfrac{1}{a}$	a^{-1}

$\dfrac{1}{a^2}$	a^{-2}
$\dfrac{1}{a^3}$	a^{-3}

And so on. The bottom box answers the question the homework is asking.

4 OUT OF ORDER: The trouble with BODMAS

Try it yourself – page 97

1. $20 - 4 \times 2 = 20 - 8 = 12$
2. $16 \div 2 + 6 = 8 + 6 = 14$
3. $16 \div (2 + 6) = 16 \div 8 = 2$
4. $2 \times 5^2 = 2 \times 25 = 50$
5. $(2 \times 5)^2 = 10^2 = 100$

Three numbers to make 6 – page 111

There are other ways; these are just examples.

(1	+	1	+	1)!	= 6
2	+	2	+	2	= 6
(3	+	3	–	3)!	= 6
(4	–	(4	÷	4))!	= 6
5	+	5	÷	5	= 6
6	+	6	–	6	= 6
7	–	7	÷	7	= 6

$$8 \quad -\sqrt(\sqrt(8 \quad + \quad 8)) \quad = 6$$
$$(\sqrt9 \quad + \quad \sqrt9 \quad - \quad \sqrt9)! \quad = 6$$

It feels utterly impossible to make 6 with three 0s, until you realize that $0! = 1$. Why on earth is that true? There are a few ways to understand why. Firstly, the factorial operation represents the number of ways of ordering a group of objects. How many ways are there to order zero objects? One way: the empty shelf! It's also worth noting that to get from $n!$ to $(n-1)!$ you must divide by n:

$$4! \div 4 = 3!$$
$$3! \div 3 = 2!$$
$$2! \div 2 = 1! \qquad \text{Therefore it follows that…}$$
$$1! \div 1 = 0!$$

And by this logic, since $1!$ is 1, $0!$ is also 1. The third, less mathematically rigorous, way to prove that 0 factorial is equal to 1 is just to shout it really loud. Observe:

$0! = 1$

Either way, once we know this fact, we can make 6 with three 0s:

$$(0! \quad + \quad 0! \quad + \quad 0!)! \quad = 6$$

5 Bad maths: When Facebook meets algebra

Try it yourself: Three simultaneous equations problems – page 117

1. Form two simultaneous equations, with x and y representing the number of adult and child tickets sold, respectively.

$x + y = 400$ (number of tickets sold)
$10x + 8y = 3900$ (the value of the ticket sales)

Next you could multiply the top equation by 10 and then subtract the bottom equation:

$10x + 10y = 4000$
$10x + 8y = 3900$

$2y = 100$
$y = 50$

50 child tickets were sold, and therefore the other 350 tickets sold were adult tickets. Note: there is a quicker, more ingenious way to solve this problem with no algebra required. If all tickets were sold to adults, the venue's income would be £4000. But in actuality it is £3900, i.e. £100 less. Every time the venue sells a child ticket they essentially 'lose' £2, and the venue has 'lost' £100 overall, compared with their maximum possible income. £100 ÷ £2 = 50, so 50 child tickets must have been sold.

2. Before we begin, note that the question requires speeds but the information we are given is in terms of distances and times. So what we really need to know is that the speed walking the 'right way' is 4m/s (100 ÷ 25) and the speed walking the 'wrong way' is 2m/s (100 ÷ 50). Now, the speed walking the 'right' way is the sum of the walking speed and escalator speed, but the speed walking the 'wrong' way is the difference between the walking speed and escalator speed. So with x and y representing the walking speed and escalator speed respectively:

$x + y = 4$ (walking 'forwards')
$x - y = 2$ (walking 'backwards')

Add the two lines together:

$2x = 6$
$x = 3$

The walking speed is 3m/s and the escalator speed is 1m/s.

3. This is my all-time favourite simultaneous equations problem (praise indeed!). I have no idea where it first came from, but an ex-student emailed it to me with a suggestive winky emoji attached.

Let x represent the mother's age at the *start* of the problem, and y represent her child's age at the same time. Then the two equations we need are as follows:

$x - y = 21$ (A mother is 21 years older than her child)
$x + 6 = 5(y + 6)$ (In 6 years' time the mother will be exactly 5 times as old as the child)

Simplifying the bottom expression and rearranging it similarly to the first:

$x + 6 = 5y + 30$, therefore…
$x - 5y = 24$

Now once more putting the two equations together, and subtracting…

$x - y = 21$
$x - 5y = 24$

$4y = -3$
$y = -0.75$

This means that the child is minus 9 months old, which we might consider to be 9 months before the child is born. Where is Dad at this point? I hope I don't have to explain that part.

The simultaneous equations magic trick – page 118

Here's how I did it; you may have a better way. Set out the two starting equations in the most general way possible:

$$ax + (a + b)y = a + 2b$$
$$cx + (c + d)y = c + 2d$$

Here a, b, c and d could represent any constants (numbers) at all. Now to get to the point where I can eliminate by subtracting, I'm going to multiply the top line by c and the bottom line by a, to give the following:

$$acx + (ac + bc)y = ac + 2bc$$
$$acx + (ac + ad)y = ac + 2ad$$

Now subtract bottom from top:

$$(ac + bc - ac - ad)y = ac + 2bc - ac - 2ad$$

Which simplifies to:

$$(bc - ad)y = 2bc - 2ad$$

or

$$(bc - ad)y = 2(bc - ad)$$

Dividing both sides by $(bc - ad)$ leaves us with $y = 2$. Now we can substitute back into one of the original equations, but with $y = 2$:

$$ax + (a + b)y = a + 2b$$
$$ax + 2(a + b) = a + 2b$$
$$ax + 2a + 2b = a + 2b$$
$$ax + 2a = a$$
$$ax = -a$$
$$x = -1$$

The solution to any such pair of simultaneous equations will always be: x = −1, y = 2.

The Cbeebies equations – page 125

Only the top two lines give us anything to work with, so let's start with:

$a + h + t = 26$ (A)
$2t - h = 20$ (B)

Adding the two lines together will cancel out the h terms:

$a + 3t = 46$ (C)

Now if we let the tree emoji stand for any value we want it to, we can find corresponding values for a and h:

$h = 2t - 20$ (from B, above)
$a = 46 - 3t$ (from C, above)

So if you want the tree to represent 5, then $h = -10$ and $a = 31$. And if you want the tree to represent 10, then $h = 0$ and $a = 16$. You can check these, or any other solution set, by substituting back into the original equations and you'll find that they work.

If apple, tree and house can stand in for any values we want them to, does that mean the bottom line of the equation can be anything we want it to as well? Not quite. Let's explore the potential values the bottom line can take.

$at - a = ?$ Which could be rewritten as…
$a(t - 1) = ?$ But we have a definition for a in terms of t above…
$(46 - 3t)(t - 1) = ?$

This is a quadratic function in *t*, meaning in this case that it has a maximum value. If you know a little about quadratic functions you could go on to find out that the maximum value the bottom function could take is 154.083, and this happens when the tree emoji is worth 8.167.

The homework grid – page 131

5	+	7	−	6	= **6**
+		−		×	
8	−	4	×	2	= **8**
−		×		÷	
9	×	1	÷	3	= **3**
=		=		=	
4		**3**		**4**	

Look-and-say sequence – page 141

1, 11, 21, 1211, 111221, 312211, …

Why would there never be a 4 in this sequence, no matter how long it goes on for? It's probably best to imagine there was a 4, then consider the consequences. The sequence doesn't start with any 4s, so the only way a 4 can first appear is if it is the count of how many of a certain number in the previous term. So, if there was a '42' somewhere amongst a term, that would mean the previous term contained four 2s in a row: '2222'

'2222' when spoken aloud according to the say-what-you-see rule, is 'two twos, two twos'. But you would never say that, you would say 'four twos', and then we're back where we started, at '42'. There's no way to create a situation in which a 4 would occur.

6 GET INTO SHAPE: Genius geometry problems

The Henk Reuling question – page 167

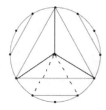

Here's the 'long way' to get the solution to the above problem. Note that the solid black lines that I've added split the design into three identical sectors. So finding the shaded/unshaded ratio of one of these sections will be equivalent to finding the ratio for the whole circle. This method makes use of the formulae for the area of a triangle and area of a circle sector.

Area of each white triangle: $\frac{1}{2}r^2\sin(\frac{2\pi}{3}) = \frac{\sqrt{3}}{4}r^2$ (A)

Area of each small white sector: $\frac{1}{2}r^2(\frac{\pi}{3}) - \frac{1}{2}r^2\sin(\frac{\pi}{3}) = \frac{\pi}{6}r^2 - \frac{\sqrt{3}}{4}r^2$ (B)

Area of one third of the whole circle: $\frac{\pi}{3}r^2$ (C)

Area of one shaded section (C – A – B): $\frac{\pi}{3}r^2 - \frac{\sqrt{3}}{4}r^2 - (\frac{\pi}{6}r^2 - \frac{\sqrt{3}}{4}r^2) = \frac{\pi}{6}r^2$

So the total shaded area is three lots of $\frac{\pi}{6}r^2 = \frac{\pi}{2}r^2$, in other words half of the circle.

CONCLUSION

Miracle sudoku – page 185

Here's a miracle sudoku to have a go at. This is an easier one because you have four whole digits to get you started (!). Once again, it's an anti-knight, anti-king, non-consecutive sudoku.

Anti-knight: Any two cells separated by a knight's move in chess cannot contain the same digit.

Anti-king: Any two cells separated by a king's move in chess cannot contain the same digit.

Non-consecutive: Any two orthogonally adjacent cells cannot contain consecutive digits. ('Orthogonally adjacent' means neighbouring squares that share an edge, i.e. not diagonally neighbouring squares.)

Oh, and normal sudoku rules also apply. The answer is on my website – good luck!

		1		2				
								2
								5

ACKNOWLEDGEMENTS

Firstly, thank you to Jessamine, Edwin and Juno for being excellent company. I love you very much. Thanks to Ed Faulkner at Atlantic for helping shape the book with his advice throughout, and to Rob Eastaway for the original introduction. Thanks to Tim Harford and Alex Bellos for their kind comments on the first draft.

Thank you to Ed Southall, Catriona Agg, Kit Yates, Claire Longmoor, Chika and Nma Ofili, Mary Ellis, Gracie Cunningham and Simon Anthony for giving up their time to be interviewed. Thanks also to Ben Sparks (ideas for several parts of this book came from one of his excellent 'maths magic' talks), James Tanton, Graham Cumming, Chris Smith and Alon Amit. The Cover was designed by Nathan Burton; Hana Ayoob did the excellent illustrations - check out her work.

Diego Rattaggi introduced me to *Potenz vor Punkt vor Strich*; Martin Noon showed me the simultaneous equations magic trick; the cable cars question comes directly from Professor Ian Stewart who gave the kind permission for its use; Kit Yates found the bad BODMAS quiz question; Katie Steckles showed me the half-dogs simultaneous equations question; Clare Wallace pinched the emoji clock monstrosity from her mam's Facebook; Benjamin Leis showed me the emoji fruit question with the 100-digit solution; Barry Dolan showed me the *I'm a Celebrity* question. (Barry also read through an early draft, and although his contribution was largely writing 'nice' next to every number 69, he also gave me some very useful feedback.)

I discovered Claire Longmoor as the creator of the viral orchestra question via aperiodical.com, an excellent recreational maths website. Lots of ideas from the BODMAS chapter come from a TES podcast by Jo Morgan and Craig Barton, and Colin Foster should be credited for the 'always subtract first' idea. Enormous thanks to Graham Dury at *Viz* for digging out the 'missing pound' cartoon, and for all the maths/science references he's sneaked in over the years.

Thank you to all my colleagues and students past and present at Barton Peveril Sixth Form College for endless mathematical inspiration, especially Matt Arnold and Paul Green who helped me work out the success probability for the calculator trick.

All of the below worked on the book in some way. I haven't met nearly all of them, but without their work you wouldn't hold the book in your hands right now. Keep buying books, and keep supporting independent bookshops and publishers.

Mairi Sutherland; Copy editor; Rich Carr, Typesetter; Emma Heyworth-Dunn, Managing Editor; Kate Ballard, Senior Editor; Alan Craig, Production Director; Niccolò De Bianchi, Production Manager; Karen Duffy, Head of Campaigns; Jamie Forrest, Marketing Campaigns Director; Clive Kintoff, Sales Director; Patrick Hunter, UK Key Account Manager; Gemma Davis, Head of International Sales; Isabel Bogod, Sales Executive; Alice Latham, Rights Director.